Technology Shocks

Contributions to Management Science

H. Dyckhoff/U. Finke
Cutting and Packing in Production and Distribution
1992. ISBN 3-7908-0630-7

R. Flavell (Ed.)
Modelling Reality and Personal Modelling
1993. ISBN 3-7908-0682-X

M. Hofmann/M. List (Eds.)
Psychoanalysis and Management
1994. ISBN 3-7908-0795-8

R.L. D'Ecclesia/S. A. Zenios (Eds.)
Operations Research Models in Quantitative Finance
1994. ISBN 3-7908-0803-2

M.S. Catalani/G.F. Clerico
Decision Making Structures
1996. ISBN 3-7908-0895-4

M. Bertocchi/E. Cavalli/S. Komlósi (Eds.)
Modelling Techniques for Financial Markets and Bank Management
1996. ISBN 3-7908-0928-4

H. Herbst
Business Rule-Oriented Conceptual Modeling
1997. ISBN 3-7908-1004-5

C. Zopounidis (Ed.)
New Operational Approaches for Financial Modelling
1997. ISBN 3-7908-1043-6

K. Zwerina
Discrete Choice Experiments in Marketing
1997. ISBN 3-7908-1045-2

G. Marseguerra
Corporate Financial Decisions and Market Value
1998. ISBN 3-7908-1047-9

WHU Koblenz – Otto Beisheim Graduate School of Management (Ed.)
Structure and Dynamics of the German Mittelstand
1999. ISBN 3-7908-1165-3

A. Scholl
Balancing and Sequencing of Assembly Lines
1999. ISBN 3-7908-1180-7

E. Canestrelli (Ed.)
Current Topics in Quantitative Finance
1999. ISBN 3-7908-1231-5

W. Bühler/H. Hax/R. Schmidt (Eds.)
Empirical Research on the German Capital Market
1999. ISBN 3-7908-1193-9

M. Bonilla/T. Casasus/R. Sala (Eds.)
Financial Modelling
2000. ISBN 3-7908-2282-X

S. Sulzmaier
Consumer-Oriented Business Design
2001. ISBN 3-7908-1366-4

C. Zopounidis (Ed.)
New Trends in Banking Management
2002. ISBN 3-7908-1488-1

U. Dorndorf
Project Scheduling with Time Windows
2002. ISBN 3-7908-1516-0

B. Rapp/P. Jackson (Eds.)
Organisation and Work Beyond 2000
2003. ISBN 3-7908-1528-4

M. Grossmann
Entrepreneurship in Biotechnology
2003. ISBN 3-7908-0033-3

Heinrich M. Arnold

Technology Shocks

Origins, Managerial Responses, and Firm Performance

With 34 Figures
and 77 Tables

Physica-Verlag
A Springer-Verlag Company

Series Editors
Werner A. Müller
Martina Bihn

Author

Dr. Heinrich M. Arnold
University of Munich
Munich School of Management
Institute for Innovation Research and
Technology Management (INNO-tec)
Kaulbachstraße 45
80539 Munich
Germany
heinrich.arnold@stanfordalumni.org
arnold@bwl.uni-muenchen.de

ISSN 1431-1941
ISBN 3-7908-0051-1 Physica-Verlag Heidelberg New York

Cataloging-in-Publication Data applied for
A catalog record for this book is available from the Library of Congress.

Bibliographic information published by Die Deutsche Bibliothek
Die Deutsche Bibliothek lists this publication in the Deutsche Nationalbibliografie; detailed bibliographic data is available in the Internet at <http://dnb.ddb.de>.
Zugl.: München Univ., Diss., 2002

Physica-Verlag Heidelberg New York
a member of BertelsmannSpringer Science+Business Media GmbH

http://www.springer.de

Softcover design: Erich Kirchner, Heidelberg

SPIN 10901267 88/3130/DK-5 4 3 2 1 0 – Printed on acid-free paper

Foreword

Research on innovation processes and their management has become an important field in economics and business administration. Given the complexity of managerial tasks in real-world innovation projects, it is no wonder that practitioners are turning to researchers for answers on how to manage their corporations in times when technical change is threatening to destroy established positions of market dominance.

One of the biggest challenges managers are facing is how to respond to technical change. It is by now one of the central tenets of the innovation literature that established firms tend to fail in the face of drastic technical change (or technology *shocks* – as Heinrich Arnold refers to them). This is a puzzling phenomenon, given the benefit of experience, resources and stability that an established player often enjoys, when compared to newly entering firms and start-ups. This observation has been made repeatedly, mostly in very detailed case studies seeking to identify the causes for the failure of the established firms in a specific context.

At the outset, there are three important questions in this field of research. First, to what extent can we generalize from the case studies that have been produced and that document the failure of established firms when they encounter particular types of technical change? After all, the selection of case studies follows the logic of picking an arena which is likely to make observing a phenomenon simple. Typically, this approach does not lead to results with high external validity. Second, what are the reasons for the observed outcomes, assuming that there is a general pattern, whatever it may be? And third, what can be done about it? Which elements of the innovation process can be influenced in times of radical technical change by prudent managerial practice?

In his dissertation, Heinrich Arnold attempts to answer some of these questions. He starts by developing a refined conceptual framework which is then used to guide his analysis of managerial responses to three consecutive technology shocks in the machine tool industry. Arnold uses quantitative econometric evidence and qualitative information from a large number of interviews held in the industry. His study puts together a very impressive dataset on the machine tool industry. His findings clearly indicate that managerial responses have an impact on the performance of machine tool firms. In particular, Arnold identifies the degree of innovation in machine architecture and adjustments in the firms' business models as key determinants of performance. He also identifies and discusses a number of other types of managerial responses that are positively associated with an increased chance of survival in the face of drastically changing technologies. These results are important, because they contribute to a deeper understanding of technical change and of suitable managerial (re)actions to such change.

This book is the product of more than three years of painstaking research, leading to a doctoral degree at the Ludwig-Maximilian-University of Munich. I am certain that other attempts to solve the puzzle discussed here will follow. Heinrich Arnold's study is a remarkable contribution to the field – I am sure that these results will find the attention of practitioners and researchers alike.

Munich, January 2003 Dietmar Harhoff

Dedication

This thesis is devoted to all people who supported me prior to and throughout the process of creating it - first of all my parents Luise and Heinz Arnold and my brother Martin. As with most academic studies, there are a considerable number of supporters without whom the outcome would have been of less quality or maybe none at all. This is why I would like to mention them here and express my thanks.

The list starts with Prof. Rainer Friedrich of the Institute for Fluid Dynamics at the Munich University of Technology, who recommended me for a grant by the Bavarian Ministry of Education and Science in 1993 which provided sponsoring for particularly gifted students in extraordinary academic programs. This paved my way to Stanford University. I would like to continue with the late Prof. Daniel Bershader of Stanford University's Department of Aeronautics and Astronautics, who strongly supported my admission to Stanford's interdisciplinary program 'Master of Science in Engineering' in 1994. My interest in innovation research and technology management stems from this curriculum.

Many thanks to my friends Daniel Kraus and Markus Schwager, both graduates of the Munich School of Business at the Ludwig-Maximilians University of Munich, who made me aware of the newly-founded Institute for Innovations Research and Technology Management (Innotec) headed by Prof. Stephan Schrader.

Prof. Dietmar Harhoff, the current director of the institute, became my research supervisor. I am very indebted to him for accepting me as one of his Ph.D. students and for his continued guidance. Prof. Harhoff and his Innotec team - Georg Altmann, Prof. Nik Franke, Dr. Marc Gruber, Dr. Joachim Henkel, Claudia Küpper (my long-time office mate), Markus Reitzig, Carolin Socher, Christian Tausend, and Rosemarie Wilcox - provided the environment and motivation for my academic studies and supported me with immense practical help. Prof. Harhoff's transatlantic contacts at MIT in Boston resulted in the opportunity of obtaining valuable input from the Sloan School of Management, in particular from Prof. James Utterback. Mercer Management Consulting sponsored my participation in the 'Entrepreneurship Development Program' at the latter.

It was a great coincidence that the first graduate curriculum for Ph.D. candidates for the 'Master of Business Research' (MBR) took off in November 1999 at the same time as the beginning of my almost two-and-a-half year long affiliation with the Munich school of business. This program was led by professors Dietmar Harhoff, Hans-Ulrich Küpper and Eberhard Witte, who put all of us Ph.D. candidates of the same year together in one class - probably the first time in the school's history. The program brought me in contact with a group of people with rather diverse talents; some of them became my friends. Markus Frühauf and Marc Rodt, who accompanied me all these years, were in this first MBR class of 1999.

I am grateful to Prof. Küpper for being the second supervisor. I am indebted for the support I received from Julie Hsieh during the initial phase of this work and Cora Laugs and Lisa Peters during the finishing phase to turn my German-flavored English into an English that tries not to give away the author's nationality.

For the great ideas on analytical statistics, I would like to sincerely thank Prof. Harhoff, Wibke Wirth, Siemens AG, and Katrin Hegendoerfer, Psychological Institute of the University of Munich. The data collection in the telecommunication infrastructure industry was made possible and generously supported by Mr. Dittberner of Dittberner Associates, who even invited me to stay at his house in Washington while I collected data from his company. I would like to thank Dr. Heinz Drink, founder of the consultancy UBM, Klaus Leinemann, independent consultant, Prof. Tom Peisl, University of Applied Sciences in Munich, Gerhard Baur, Siemens AG, Nik Skaperta and Hans-Eugen Binder, former heads of the research department of Siemens switching technology for making available both their mental and paperwork archives to me. I would like to thank Dr. Kai Lamottke, managing director of Bicoll Biotechnologies, for the support and possibility of applying and transferring some of the research findings to the setting of a biotech start-up. I would particularly like to thank Mr. Dzems Bruvelis, owner and managing director of UBM, who was a real mentor during the process of the generation of ideas as well as data collection and data collection within the machine tool industry. Without his commitment to accompanying me to several of the interviews or his intuition and willingness in supporting my quest for information and contacts, a quantitative analysis would not have been possible. Many of the ideas contained in this research are based on discussions held with Mr. Dzems Bruvelis.

I would also like to thank my friends and roommates Chrissi Ohlwärter and Dr. Andi Hutter, who provided an understanding, caring, and motivating environment during the last few years. In addition, I would like to thank Andi's uncle Werner Hutter, who - as an expert in switching technology - opened the door for me to telecommunications infrastructure technology and put me in a position to expand my research into this area.

Mercer Management Consulting supported me by consenting to an exceptionally long leave of absence as well as providing me with a sponsorship. I would like to thank Ghyhaelaine Young-Wales of the Harvard School of Design for tactical advice regarding the distribution and communication of research results.

And lastly, I would like to thank Victoria Portarescul, Tom Glück, Stefan Hollinger, and Bernhard Lohnert for their continued interest in what I was doing (including some time-critical delivery of documents, data input, formatting and good spirits). Thanks to Victoria for agreeing to become my wife during this research: the birth of our son Maximilian Martin on January 1st of this year motivated me to quickly finish up the final formatting process.

During the course of this thesis, I will mention other contributors without whose information some of the chapters would have been less informative.

Munich, March 2003 Heinrich Martin Arnold

Motivation and mode of use

Radical technological changes - or more intuitively called 'technology shocks' - are frequent causes for changes in the competitive structure of industries. Market leaders lose their dominant positions, new entrants appear; in some cases the borders of industries are redefined, in some instances, former market leaders disappear entirely.

The often quoted example of the semiconductor industry illustrates this phenomenon well. The largest US manufacturers of the early 60s (e.g. Fairchild, General Instruments, General Electrics, RCA, Sprague, Philco, Transitron, Raytheon) were unable to maintain their market positions. The emergence of new technologies and products (Integrated Circuits, VLSI, Submicrons) was the reason why none of them were able to remain among the industry's leading ten firms - with the exception of Texas Instruments and Motorola.

Recent research[1] has focused on the patterns of failures of incumbents and on the determinants of success among new entrants in fast-moving industries. This thesis builds on observations in industries which are also subject to technological change but are developing more slowly. The focus is on managerial actions that lead established firms successfully through radical technological change.

The impact of such factors regarded as important by managers that have not yet received much attention by scientists are also included in the research. This thesis therefore starts by reviewing the state of the art. Research is aimed at creating a framework that can be used to sufficiently describe the phenomenon of a 'technology shock' (chapter 1), and at identifying potential success factors from literature. Three groups of hypotheses are built from the following success factors: First, characteristics

---

1 C.f. Clayton Christensen 'The Innovator's Dilemma - When new technologies cause great companies to fail'. He refers to the disk drive industry, where only IBM from 17 leading companies in 1976 (e.g. Diablo, Ampex, Memorex, EMM, Control Data) was able to remain in the market until 1995. Technology shocks in this context were new technologies for the read-write head (ferro-oxide, thin film, magneto resistive technologies), also for encoding, servo-systems, motors and interfaces.

of companies that evolve over a longer period of time and are already present prior to the technology shock, second, strategic decisions and third, operational decisions made by the management in order to react to the shock (chapter 2). Some of the factors have already been analyzed statistically in the context of technological change in the past. These are used to construct and test a base model for successfully managing technology shocks in chapter 6. Factors that have been suggested conceptually, through case studies, or empirically in the wider context of innovation are taken into consideration to improve the base model.

Chapter 3 illustrates the methodology applied to this research while highlighting potential 'research gaps'. The approach is tested in the machine tool industry[2]. Data has been collected following several rounds of pre-tests. The latter have been used in order to modify and complete the list of questions and to optimize the data collection process. Identifying contacts that were willing to participate in the study was a time-consuming effort and required extensive 'networking'. Almost 300 machine tool manufacturers[3] were approached.

To identify technology shocks, the recent history of the machine tool industry has been reconstructed from literature and around 100 interviews with industry experts and executives who witnessed the past decades in the industry (chapter 4). The development of the dimensions that indicate technology shocks (i.e. machine architecture, companies' competencies, and performance parameters) has been examined. In addition, current technological trends are analyzed in order to identify potential shocks of the future and to indicate the practical relevance of this research.

---

[2] The approach has also been tested in the telecommunication infrastructure industry (in this case switching technology: transition from mechanical to electromechanical to electronic to IP-switched). In addition, downstream industries - e.g. telecom operators, capital goods manufacturers - could be subject of further research in order to extend the external validity of the success factors identified. Several companies of these industries have confirmed the necessity of and expressed their interest to participate in such research. As for the generation of hypotheses, the telecommunications industry has been included in this research. Findings from the machine tool industry and telecommunication infrastructure industry have been transferred to the biotech industry by means of an experiment: the research findings were translated into 'managerial recommendations' that became part of the strategic management of a biotech start-up company. Results from this experiment and from the telecommunication infrastructure industry are not included in this thesis. Trying to discuss and evaluate all information that was collected in the process of this research would be at the expense of the focus and readability of this work.

[3] and 10 telecommunication infrastructure manufacturers

Technology shocks are accompanied by shifts in market positioning and failures of some of the leading companies. It is expected that success and failure are caused by company attributes that have evolved in the companies' histories and are the results of a long-term managerial process as well as strategic and operational decisions in reaction to the technology shocks. Top managers from 58 machine tool companies were willing to supply data in response to a detailed 10-page question catalogue. An analysis of the three groups of hypotheses defined in chapter 2 discloses the factors that differentiate successful companies from those that failed (chapter 5). In addition, managers were asked to rate the importance of the different success factors. Their views are used to focus the analysis on critical issues.

Figure 0.1: Overview of research process and thesis structure

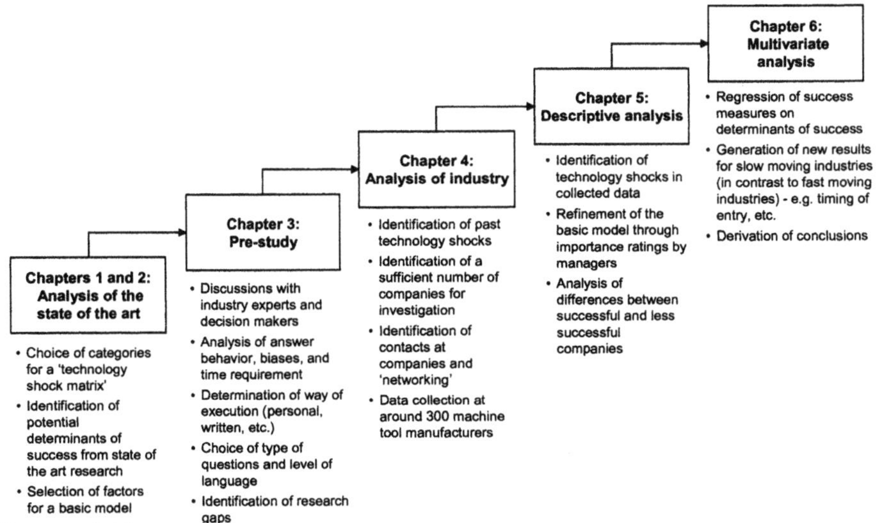

Source: own depiction.

The last step is a multivariate analysis of the duration of survival and failure frequency in chapter 6. The result is a model describing how companies can successfully navigate through technology shocks.

This thesis makes use of very diverse approaches. It takes a business history perspective and a case-related view to describe and characterize the industry serving as the research object. It examines literature to summarize the state of the art and 'research' gaps indicating the direction for research. It uses prognostic tools in order to identify potential future technology shocks and illustrate the practical applicability and relevance of the research results. It makes use of descriptive and analytical econometric statistics to identify significant influences among factors that potentially influence the successful management of technology shocks. An overview of the research process and how it corresponds to the structure of this thesis is given in figure 0.1.

Contents

1 Frameworks to capture technology shocks: an attempt to integrate current streams of thought

1.1 Introduction

Former studies have identified effects of so-called radical technological change - which should be referred to as 'technology shocks' - as being critical to company success and eventually determining survival in the corporate world. Many approaches have been taken to characterize technological change and its effect on companies' competitiveness. This chapter summarizes current streams of thought on this topic, which are used to select the most appropriate forms of categorization. A framework for identifying severe technology shocks can then be created.

In order to find a structure for technology shocks along the timeline, the similarity of technology shocks to the product life cycle concept (PLC) is being investigated. It becomes evident that technology shocks follow a four-stage pattern similar to that of products.

The chapter develops a scheme to characterize and categorize technological change as the basis of further analysis. The findings of this chapter may be applied two-fold: first, as a screening tool to assess the severity of technological change, and, second, as a generic map laying out the development of the competitive environment over time.

1.1.1 Introducing technology shocks

Scholars and practitioners with different approaches and perspectives have examined technology as a major determinant of competitive advantage. A number of different terms are related to or contain elements of the same phenomenon. In a broader context, Schumpeter described innovation as the creation of a new production function (Schumpeter, 1939) consisting of several factors, one of which is technology. Subsequent authors followed this idea. Among them was Hollander, who defined technical change as the degree to which a linear homogeneous production function[1] is changed (Hollander, 1964). Several authors have dealt with major innovations (Cooper et al., 1992), technical discontinuities (Tushman et al., 1986), (radical) technological changes (Chanaron, 1999; Gebhart, 1995; Goldhar, 1994; Martino, 1999), recently also termed disruptive technologies (Bower et al., 1995; Christensen, 1997), or similar concepts. Most of these contributions are named 'technological change' in major searchable literature databases[2].

The concepts provided in current publications cover a range of different aspects of the same phenomenon. Their potentially endangering effects on companies is quite similar. There is a need for consistent terminology. It is characteristic of the literature to refer to the phenomenon in terms of two regimes (usually regimes characterized by an old and a new technology, but also by classical and new competence, by old and new markets, etc.), where one regime replaces the other. Change can be observed in the characteristic parameters of the respective regimes. Such characteristic parameters are examples of relevant competence, successful sales strategies, customer demands, etc., which differ in the old and the new regimes.

In physical fluid dynamics, the rapid adaptation of one regime[3] to another causing a discontinuous change in the main parameters is called 'shock'. In order to introduce consistent terminology the term 'technology shock' will be used here to encompass and substitute a number of concepts. The terminology and the systematics in natural sciences have often enriched social sciences. In this case, it appears beneficial to borrow from physics.

The term 'technology shock' has only been used in a different context so far: in economic research when modeling business cycles using the real business cycle method (RBC) (Eichenbaum, 1991; Rao, 1997). R&D spending and patent applications in the industry are among the empirical indicators for technology shocks in economic research (Shea, 1998).

Several suggestions for identifying and describing technology shocks at the firm level are made in the following section: technology shocks can appear in the form of radical or architectural innovations and in this way differ from incremental and modular ones. The extent to which a technol-

---

1 Cobb-Douglas aggregate production function applied to US data.

2 ABI/Inform and BLISS databases as provided by the Universitätsbibliothek München - available at http://www.ub.uni-muenchen.de/Datenbanken.htm are the sources for this analysis.

3 A regime is equivalent to a certain state of a fluid. It is characterized by thermodynamic parameters such as molecular weight of the substance, pressure, density, and velocity.

ogy shock has occurred can be described using technical parameters and properties of the product or by measuring the impact of technological change on a company's competence. The first goal of this chapter is to find the most suitable dimensions along which the technology shocks can be identified, characterized and measured.

1.1.2 Characterizing technology shocks at the firm level

Technological factors, along with competitive, legal, and political factors shape an industry's environmental condition. Organizations are dependent on their environment for survival (Pfeffer et al., 1978).

The evolution of an industry's environment is not usually a gradual one. Abrupt technological changes disrupt the environment[4] (Mensch, 1979) with severe consequences for the organizations. The evolution of technology occurs in periods of incremental change that are impacted by technological breakthroughs (Loch et al., 1999). These breakthroughs can either enhance or destroy the competence of firms (Tushman et al., 1986), thus creating 'unmistakable challenges for established firms' (Henderson et al., 1990). Tushman and Anderson describe major technological changes in products and processes. They conclude that product discontinuities lead either to the emergence of new product classes (e.g. airlines, automobiles, plain paper copiers), to product substitution (e.g. transistors vs. vacuum tubes, diesel vs. steam locomotives) or to fundamental product improvements (e.g. jets vs. turbojets, LSI vs. VLSI semiconductor technology). The same phenomenon can be applied to the processes. Discontinuities in this case lead either to process substitution (e.g. mechanical ice-making vs. natural ice-harvesting, thermal vs. catalytic cracking in crude oil refining, artificial vs. natural gems) or to process innovations that radically enhance industry-specific performance (e.g. increase of glass output through the introduction of float glass, increase of cost-efficiency through the introduction of mini steel mills) (Tushman et al., 1986). 'The emergence of a new industry based on major product innovation (such as the electronic calculator industry in the 1960s) often poses the threat of substitution to companies with a base in a more established industry.' (Cooper et al., 1992). Technology shocks deeply affect the organizational state and the economic performance of the investigated company or industry (Henderson, 1993), potentially leading to failure (Cooper et al., 1992). In failing to innovate, an existing business may be taken over by a competitor who does innovate.

[4] Mensch differentiated between basic and incremental innovation.

Thus an important incentive for firms to attempt risky technological innovation is to survive in the face of effective competition (Hill et al., 1980).

The second goal of this chapter is to find out if there is a relationship between the evolution of competitive environments and strategic issues which companies face because of technology shocks.

A technology shock expresses itself in significant changes in sales volume or value due to the shift from old to new technology (Cooper et al., 1992).

The above-mentioned sources indicate that the occurrence of technology shocks is a widely discussed and significant phenomenon. Peculiarly, the literature's main focus thus far has been on discussing the occurrence of radical technological change and on explaining its destructive effects. In this thesis technology shocks triggered by external factors are regarded as given. The focus of the analysis in this work is rather on the company internal factors that management can influence to pass successfully through technology shocks.

1.2 Frameworks

Innovation and technological change need to be operationalized and measured in order to arrive at an empirical assessment and to provide tools and guidelines for practitioners. Since Schumpeter's work in the 1930s and 1940s (Schumpeter, 1934, 1939, 1942) much effort has been put into differentiating between incremental or continuous and basic, radical, or discontinuous innovations (Anderson et al., 1990; Baba, 1989; Ehrnberg et al., 1997; Glasmeier, 1991; Henderson et al., 1990). Incremental technological progress is a gradual, cumulative process stimulated by the prospect of economic returns until it is disrupted by a major advance. Major advances typically present a discontinuous improvement of the price/ performance ratio as compared to existing technologies. 'Major technological innovations represent such technical advance that no increase in scale, efficiency or design can make older technologies competitive with the new technology.' (Anderson et al., 1990; Baba, 1989; Ehrnberg et al., 1997; Glasmeier, 1991; Henderson et al., 1990; Tushman et al., 1986). Several empirical studies follow this approach of interviewing technical experts about the importance of innovations. In addition, patent counts are often used to approximate the innovative output of firms or industries. Cohen and Levin provide a summary of these approaches (Cohen et al., 1989). An overview of frameworks to categorize technology shocks is presented in the following.

1.2.1 Innovation of product properties

Henderson has introduced two dimensions to characterize product properties and differentiate between incremental and radical innovation (Henderson et al., 1990). By projecting the change of product[5] properties onto two dimensions, that can be roughly identified as 'change of correlation between components' and 'change of core concepts', she constructed a matrix containing radical, architectural, modular, and incremental innovations (see appendix A.1.1). Radical and incremental innovations mark the extreme points along the diagonal of both dimensions. With regard to design aspects, radical innovation establishes a new design which expresses itself in new core concepts in addition to the links between components and concepts. Incremental innovation, on the other hand, refines and reinforces an established design. Some components are improved upon, but the underlying design remains untouched. Companies are familiar with incremental innovation because it occurs frequently. Component change can usually be managed as a routine.

Henderson emphasizes two further types of innovation: modular and architectural[6] innovation in her framework. Modular innovation changes core concepts but not the overall product architecture. Architectural innovation, on the other hand, entails a new overall design but leaves the components and the core concepts used in the design unchanged.

It is obvious that radical innovations create life-threatening challenges for companies (Cooper et al., 1976). Architectural innovation often creates problems for established firms less conspicuously (Henderson et al., 1990). It is a more subtle phenomenon since it affects a company's communication channels, information filters and problem-solving strategies, in which architectural knowledge is stored - sometimes implicitly (von Hippel, 1994). Established organizations require significant time and resources to identify a particular innovation as architectural or they might misinterpret the nature of a threat by relying on old frameworks of experience.

Afuah and Bahram have taken Henderson's concept one step further by adding another dimension, naming the resulting three-dimensional graph 'the hypercube of innovation' (Afuah et al., 1995). They conclude that innovations can affect players differently along the value chain. What might

---

5 When discussing products, processes and services are also being addressed.

6 Abernathy and Clark (1985) have already used the term 'architectural', although with a different meaning. They describe architectural innovation as a situation where new technology redefines both the basic configuration of products or processes and the linkages to markets and users, making existing competence obsolete.

be radical for the supplier could be an incremental innovation for the customer, for example. Afuah and Bahram differentiate between the supplier, innovator, customer and complementary innovator along the value chain (see appendix A.1.1).

1.2.2 Change in product's value proposition

Christensen differentiates sustaining technologies, which bring about improved product performance, from disruptive technologies. The contrast between disruptive and conserving technologies has already been discussed by Abernathy and Clark (Abernathy et al., 1985). Christensen deviates from the product/process-centric differentiation described above by including a product's value for its customer in his set of criteria. Thus disruptive technologies bring to the market a different value proposition than had been available previously. In other words, disruptive technologies do not necessarily compete with the older technology in precisely the same market for precisely the same customers at the beginning. At first, the performance of the disruptive technology is below customer requirements. Christensen observed what might seem paradoxical at first sight - disruptive technologies initially underperform versus established products in mainstream markets. However, they present a different value proposition to the market: 'they have other features that a few fringe - and generally new - customers value.' (Bower et al., 1995; Christensen, 1997; Christensen et al., 1996) (See appendix A.1.2). Continuous innovation improves the new technology's performance faster than the growth of most customers' requirements. As a result, the new technology becomes acceptable to the majority of customers, replacing the older technology. Trajectories of product performance over time and downward shifts at the time a new technology is introduced may represent a change in the value proposition. Christensen calls an innovation 'sustaining' when it improves a product's performance. When an innovation brings about a new value proposition that initially leads to poorer product performance, it is 'disruptive'. Christensen has since developed his concept further and refers to technologies as 'disruptive' when they enable the creation of new business models (Christensen et al., 2000).

The claim that new products need to create a higher customer value in order to induce the substitution of established products is not a new one. The literature has enumerated ample examples of how new technologies have enabled products to provide higher value for customers and to replace products based on older technology, while introducing a different value proposition (Levitt, 1960).

The examples Christensen provides - disk drives (Christensen, 1993; Christensen et al., 1996) and mechanical excavators (Christensen, 1997) - illustrate that disruptive technologies are often similar to architectural or radical innovations. But architectural innovations can also be just as sustaining in character as incremental or modular innovations. It seems that Christensen's performance measurements and Henderson's categories lie along the same lines - shifts in the performance trajectory help differentiate radical from incremental and modular innovations.

Great similarities can be found when comparing Christensen's examples of disruptive technologies with the concept of architectural innovation. Henderson did not explicitly relate architectural innovation to a downward shift in the performance trajectory as it is typical for disruptive innovations. But the two concepts can be aligned.

I suggest the following consideration. An architectural innovation could result in a lower performance of some key parameters while offering some additional features lacking in conventional products. In some instances, it is possible that no fringe customers would accept lower product performance in exchange for the other benefits that the new product provides. In this case, the product would not appear on the market until its performance has improved and it has surpassed the old product with respect to conventional key parameters as well. It might seem that this kind of innovation is sustaining in character. The resulting disruptive change in performance parameters would not appear in product catalogues or the like, but the disruptive performance characteristics can be observed in internal documents like lab reports.

1.2.3 Effects on organizations' competence

Another way to measure technology shocks is to determine the influence they have on a company's competencies. Discontinuities in technology can either enhance or destroy the competence of established organizations. Competence-enhancing discontinuities can still bring about order-of-magnitude improvements in price performance. By building on existing know-how pertaining to a particular product class, they consolidate leadership in that product class (or industry as far as process innovations are concerned), thus increasing barriers to entry. Competitive constellations of competence-enhancing discontinuities are characterized by a reduction in firm entries versus exits and a decrease in inter-firm sales variability, i.e. market shares are comparatively stable (Tushman et al., 1986). In contrast to competence-enhancing discontinuities, competence-destroying '(...) discontinuities are so fundamentally different from previously dominant technologies that the skills and knowledge base required to operate the core technology shifts' (Tushman et al., 1986). Following the idea of 'creative destruction' - as expressed by Schumpeter (1934) - competence-destroying discontinuities potentially break the existing order and threaten the position of established organizations. Barriers to entry are lowered as the technology and sometimes business knowledge of established players is rendered worthless. New firms can more easily enter previously protected markets by exploiting the new technology. They enter the market with qualitatively different skills and expertise from the start, while formerly dominant firms have to undergo a process of changing skills, abilities and expertise from prior technologies and ways of operating. New entrants take away sales from the established players while more companies enter the market than leave it. Inter-firm sales variability is high, forcing some established companies to leave the market. According to Leonard-Barton, competence can be defined as the whole set of capabilities of a company. It includes employee knowledge and skills, which are embedded in technical and managerial systems as well as values and norms (Leonard-Barton, 1992).

Abernathy and Clark used the competence view as one dimension to analyze the competitive implications of innovation and illustrate contingencies between technology and markets. The competence view is also referred to as the 'technology side' of the firm. It embraces aspects of design, organization, skills, supplier relations, capital, and experience. The second perspective is the effect of innovation on a company's relationship with markets and customers. It is also referred to as 'the customer or market side' of the firm, which includes such aspects as the relationship with the customer base, available applications, service and distribution channels, customer knowledge, and communication. The combination of the two views creates four domains of a two-dimensional matrix: architectural, revolutionary, regular innovation, and niche creation (Abernathy et al., 1985) (see appendix A.1.3). In contrast to Christensen's approach, Abernathy and Clark's categories appear to be more static and do not include technologies that develop from a niche into the mainstream market. Abernathy and Clark conclude that changes which lead to niches are built on established technical competence. The niches themselves are transitory. In contrast, Christensen's disruptive technologies originate in niches and subsequently enter the larger markets of established technologies.

In contrast to other studies Abernathy and Clark have already included a large number of different aspects to define the two dimensions of categorizing innovation. The resulting matrix becomes a broad basis for discussing the mutual influence of many aspects; but in a sense, cause and effect are mixed from the beginning. The matrix becomes a loaded tool with the ability to cover diverse phenomena, but at the expense of losing accuracy.

The examples Abernathy and Clark provide from the automotive industry are not as technologically diverse as the ones Henderson uses to illustrate her categories: most of the examples from the automotive industry would be incremental or modular in Henderson's sense, some would be architectural, but none would be truly radical. The term 'architectural innovation' has different meanings in the two studies. From a technological point of view, Henderson's categories cover a broader range of technological change and allow for a more distinct differentiation.

Although competence is defined differently in the contributions above, it has become clear in the discussion that the impact of technology on companies' competencies is a suitable dimension along which technology shocks can be categorized.

The illustration below provides an overview of the various aspects used to differentiate types of technological change.

Figure 1.1: Key aspects to characterize types of technological change

Key aspect / **Authors**	Technological advance	Customer relationship, competence	Product properties	Value chain	Value proposition/ product performance
Anderson, Tushman '86, '90 Baba '89 Glasmeier '91 Ehrnberg, Jacobsson '97 et al.	*Incremental, continuous vs. basic, radical, discontinuous*		◐		
Abernathy, Clark '85	◐	*Niche, regular, architectural, revolutionary*	◐	◐	◐
Henderson, Clark '90	●		*Incremental, modular, architectural, radical*		
Afuah, Bahram '95	●			●	
Christensen, Bower '95, '96, '97	●		●		*Sustaining vs. disruptive*

Legend: ● = Authors who included this aspect ◐ = Authors who partially included this aspect

Source: own depiction.

1.2.4 Differentiation between consumer and investment goods

One of the examples Christensen presents is the market success of off-road motorcycles from Honda, Kawasaki, and Yamaha against powerful over-the-road cycles made by Harley-Davidson and BMW. This case is not suitable for illustrating disruptive technologies. Rather than supporting his case of disruption through lower performance, this case indicates the importance of another differentiation which has not been explicitly recognized in the literature on technological change so far: the distinction between consumer goods markets and investment goods markets. While 'rational' parameters like product performance, the cost of use, etc. are decisive in the investment goods industry, the individual subjective perception of the product, called 'taste', has much more influence on purchasing decisions in the consumer market. Taste will influence the association of value with product (performance) characteristics to a high degree. Innovation is often perceived differently than technical parameters might suggest. Even class preferences and other social and cultural factors contribute to customer preferences. 'There is an economy of cultural goods, but it has a specific logic' (Bourdieu, 1984). The purchase of a car illustrates this: the status and emotions connected with an Audi A4 increases its preference over cars with comparable performance but lower prices, e.g. the Skoda Octavia (both cars are actually built on the same Volkswagen platform with quite similar technical performance). Econometricians have concluded that the value of an innovation is difficult to assess, especially when the innovation occurs in the consumer products area (Griliches, 1979). Thus, it is necessary to differentiate between consumer goods markets and investment goods markets in this context.

1.2.5 An optimized framework to identify technology shocks

A diagram is proposed to illustrate the dimensions of identifying technology shocks (figure 1.2). The segments considered technology shocks are shaded. One of the axes of the two-dimensional matrix consists of the effects technological change has on the company's core competency. The other axis expresses changes in product design and performance respectively. For investment goods and goods that yield patterns similar to investment goods performance, parameters are perceived rather objectively by the customers and correspond well to changes in the product design. When investment goods change their performance trajectory, this indicates that a change in its customer value has occurred in the way Christensen (1997) suggested.

In the case of consumer products, the technical product properties are primarily dominated by a subjective perception of performance and seem less important. The role of measurable technical performance is to a large extent replaced by a perceived performance influenced by the taste of the decision-maker. Naturally there are also investment goods where taste and image influence the purchasing decision. For example, German machinery benefits from a strong quality image that contributes to buyers' decisions. An analysis of international performance shows that a portion of this image stems from anecdotal evidence and not always from a comparison of measurable attributes (Pfeiffer et al., 1999). On the other hand, there are other examples demonstrating that purchasing decisions of consumer goods are made predominantly on the basis of price/ performance and not on the basis of taste. When customer products become commodities, taste definitely plays a subordinate role while the question of price dominates.

At this point, it has become evident that technological change or innovations can be measured along different dimensions with probably varying outcomes. By including all of the major dimensions in a 'technology shock check', it might be possible to make a more adequate judgment about the extent of technological change and its devastation for the firm.

From sections 1.2.1, 1.2.2, and 1.2.3 follows a concept of technology shock that consists of three main components: the innovation of product properties (similar to what Henderson proposed), performance trajectory, and effect on the organization's competence. The first two are to some degree interchangeable and form the technology dimension. The diagonal arrow in figure 1.2 from the lower left to the upper right corner of the matrix expresses the degree of technological change. Technological change which is being displayed in the upper right-hand corner of the matrix represents an extreme shock for the respective company, whereas changes in the lower left-hand corner would signify no shock for the company. Its effects would be quite limited. The term 'technology shock' hence describes events mapped above the diagonal in the upper right-hand corner of the matrix.

This framework for technology shocks expands on the concepts mentioned above. It introduces a differentiation between consumer goods and investment goods, aligns Henderson's and Christensen's perceptions of product performance, value proposition, and product properties, and explicitly combines the product-centric view with the competence view in one matrix.

Figure 1.2: Segmentation of technological change - dimensions of technology shocks

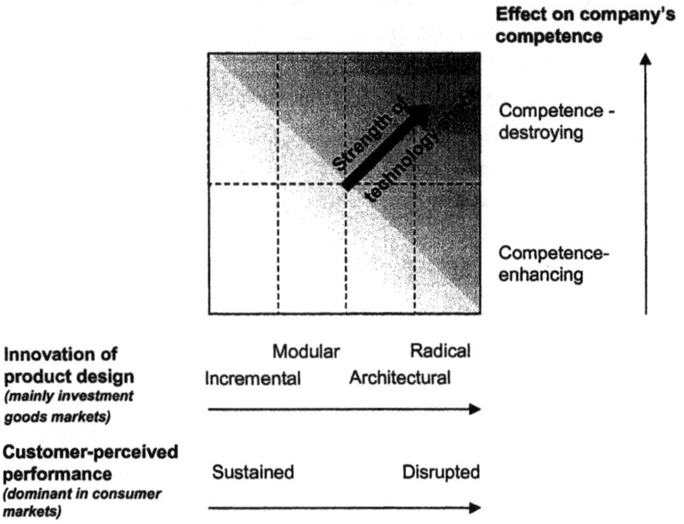

Legend: [ ] = strength of technology shock.
Source: own depiction.

1.3 Patterns of technology shocks and the product life cycle

The literature dealing with radical technological change implies that - similar to the life cycle theory (Smith et al., 1988) - the phenomena preceding and following technology shocks can be structured in stages. Klepper outlines patterns found in empirical studies supporting the product life cycle (Klepper, 1996). He also builds on other literature on technological change to support his statements. In his work, he identified four observables which hint at the similarities of technological change and the PLC: number of firms, innovative concepts and ideas, market share, and focus on product or process technologies. First, the number of firms follows a life cycle curve. There is a large number of new entrants at the beginning which decreases over time. Together with established companies, the total number of companies grows until it reaches a peak. This is followed by a decline despite the continued growth in industry output. Secondly, new entrants bring in a disproportionate share of innovation. Thus the diversity of competing versions of a product reaches a peak during the growth phase of the total number of companies. Thirdly, company market shares are volatile at the outset of a technology cycle, whereas they stabilize towards the end when industry leadership is established. Finally, the companies' focus of attention increasingly turns from product to process innovation over time.

The following sections deal with statements concerning the life cycle of products and services. The objective is to find out the extent to which the phase structure that can be extracted from the literature on technological change (and ultimately on technology shocks) indicates that technology shocks trigger a wave that follows a product life cycle pattern. The first step examines literature that explicitly deals with technological change and collects statements that indicate a stage structure. The second step examines the degree to which these findings correlate with classical product life cycle literature. The result is that technology shocks do indeed initiate a pattern similar to that of the classical product life cycle, although the latter represents a concept that was developed for consumer products.

1.3.1 Literature focusing on technological change

The different stages of technological change extracted from appropriate literature can be characterized by market introduction, rapid growth, establishment of dominant design, and cutthroat competition (Cooper et al., 1992; Henderson et al., 1990). The characteristics of the different stages triggered by technology shocks are described below.

During the **introductory stage** of a technology shock, 'the extent to which a substitution effect will occur is rarely clear. Nevertheless, managers of firms in the established threatened industry must decide how to respond to an innovation that has the potential to alter or destroy their companies' existing business' (Cooper et al., 1992; Schumpeter, 1934). Early versions of the product are often crude and expensive. Their application is limited (Cooper et al., 1976). Before a dominant design becomes established, there is much experimentation with different technical approaches and product designs (Abernathy et al., 1978). There might be even competing technologies within the same company. This situation is called 'hedging' when it is actively promoted by management[7] (Hatfield et al., 2001).

The number of firms - both new entrants and established players of the older technology - entering the field increases dramatically during the

[7] In the case of technological discontinuities in the PC industry, it was found that hedging increases the survival rate of companies and is related to higher market share.

rapid growth phase. Mansfield finds that dramatic innovation indeed has an effect on the degree of concentration in a market (Mansfield, 1983). The increase reaches its apex with the establishment of a dominant design (Christensen, 1997; Christensen et al., 1998).

A **dominant design** often emerges (and reinforces itself) in response to the opportunity to obtain economies of scale or to take advantage of externalities (David, 1985). Simplified views interpret the substitution of old through new technology through competition between old and new technology. In fact, new and different technological approaches are possible and compete with each other to substitute the older technology in many cases. One of these technologies - not necessarily the most desirable one - becomes dominant as the result of a series of rather random events (Arthur, 1988). The 'dominant design reflects the emergence of product-class standards and ends the period of technological ferment' (Tushman et al., 1986). Schewe investigated imitation strategies as an option of technology management (Schewe, 1994). One of his conclusions is that success of imitation is greater the more similar the imitation is to the actual innovation.

Although there is continued market growth, market consolidation manifests itself in the shakeout of firms during the phase of **cutthroat competition**. There is increased attention on production processes during this phase. Moreover, as experimentation with technology decreases after a dominant design has been established, the risk of an additional technology change is minor. Therefore the willingness to invest in capital-intensive methods increases, which in turn intensifies the shakeout of financially weaker companies (Abernathy et al., 1978).

Regardless of their advantages, newer technologies are not embraced by all buyers immediately - instead, a diffusion process takes place (Norton et al., 1987; Price, 1996). In their descriptive analysis of seven radical technological changes in seven industries, Cooper and Schendel (1976) found a time lag between the introduction of a new technology and the actual decline in sales of the old technology. A period of 5 to 25 years passed until the new technology overtook the old one in sales. They found that established firms usually made an effort to improve the competitiveness of the old technology while applying the new technology.

The diagram below illustrates the four stages which characterize the competitive environment and strategic issues companies are facing following a technology shock. It seems to be appropriate to refer to the phenomenon displayed below as a 'shock wave', which is triggered by technological change at the beginning and extends itself through several phases.

Figure 1.3: Four phases of a technology shock wave

Phase	**Introductory state**	**Rapid growth**	**Establishment of dominant design**	**Cutthroat competition**
Product/ process	• Preliminary product design	• Period of experimentation	• Defacto product-class standard	• Increased attention to production processes
Technology	• Often rapid growth in patents		• End of technological ferment	
Market/ competition	• Uncertainty about extent of possible substitution effect	• Rapid growth in –number of companies –market size	• Apex of growth in number of firms	• Continued market growth • Market consolidation: shakeout and reduction of number of firms
Decision level	• Strategic decision on how to respond	• Strategic decision on how to respond	• Often to obtain economies of scale or take advantage of externalities	
Investment character	• Risk and uncertainty	• Risk and uncertainty	• Reduced uncertainty	• Willingness to invest in capital intensive methods

Source: own depiction.

1.3.2 The theory of the product life cycle (PLC)

The objective of the following is to analyze contributions which explicitly describe the nature of the PLC in order to compare and contrast statements regarding technological change.

The purpose of a PLC model can be quite different depending on its group of authors (Hoffmann, 1972). One group interprets PLC as different stages of (demand-driven) market capacity (Polli et al., 1969), while another group emphasizes the change of companies' competitive behavior in the face of different sales trends of a product (Hoffmann, 1972). Both aspects also appear in the literature on radical technological change.

Investigations on technological change mainly deal with investment goods in order to document the technological parameter of change (thus implicitly eliminating the influence of subjective preferences like taste, status, etc., which in turn reduces the importance of brands[8]). In contrast to

8 Cf. the concept of psychological obsolescence in fashion industry as contrary to technical and qualitative obsolescence (a product that is still functioning is be-

this, short-lived consumer articles (both products and brands) have been the object of investigations for the PLC (Tsurumi et al., 1980): the investigations concentrate on the life cycle of a single-branded product, thus emphasizing a sub-market of a company. The total market view is only of interest when the number of competitors changes over time and thus influences the structure of products offered (Hoffmann, 1972). More recently there have been cases of applying the PLC to high-tech industries, e.g. the pharmaceutical and health care industry (Pisano et al., 1995), the mainframe computer market (Greenstein et al., 1998), the PC market (Bayus, 1998), and the overall US manufacturing industry (Lee et al., 1994). Practitioners' approaches have tried to transfer the PLC model to a model of technology cycles (Foster, 1986).

It appears that the classical PLC concept has not been applied to the entire industry: industrial goods have hardly been investigated in regard to PLC (Rink et al., 1979).

In general, all products are assumed to follow a PLC in that they enter a phase of losses towards the end of their product life and are replaced by new goods (Kotler et al., 1995). Thus Kortge et al. have proposed to take a product's progress within its expected life cycle into account when calculating prices (Kortge et al., 1994). In this way, the PLC concept can be used in business planning and appears to be superior to prognoses which are, for example, purely based on stochastic distribution models (Bollen, 1999; Gardner, 1987). Empirical research has shown that the curves of sales and revenue can have different shapes (Cox, 1967; Polli et al., 1969). The more specifically the term 'product' is defined, the better PLC characteristics can be observed. Product groups (e.g. light beer, hand-held drilling machines) follow the PLC curve more closely than product classes (e.g. beer, drilling machines) or brands/ product variations (e.g. Franziskaner Hefeweißbier, Hilti TE 10) (Siegwart et al., 1995).

1.3.3 The various phases of the PLC

The literature describes the PLC model experiencing the following phases: introduction, growth, maturity, saturation, and decline. Growth and maturity are combined into a boom phase at times. Measurables are sales, revenue, profit, and gross margin. Later contributions include the number of new competitors (Bayus, 1998). The curve is shaped either like a normal distribution curve (Gaussian bell curve), or an S-curve when the respective

ing replaced by a new product; also, a product contains parts of minor quality that wear faster) (Siegwart and Senti, 1995).

measurable is displayed cumulatively (logistic product-life curve, cumulated normal distribution) (Chrubasik et al., 1987; Siegwart et al., 1995). The five-phase scheme has also been transferred to the life cycle of enterprises (Miller et al., 1984). The various phases of product life have been amply described in the relevant standard literature (Siegwart et al., 1995). The following elaborates on just a selection of characteristic observations:

The introductory phase of a new product is characterized by losses due to initial investments. Buyers during this phase are characterized as 'innovators'.

The growth phase is characterized by the entry of imitators. Classical life cycle literature claims that scale effects already occur during the growth phase. Literature focusing on technological change does not detect large-scale effects before the third phase, during which a dominant design is established. DiMaggio and Powell found that there are a number of different organizational approaches during the early phases of a product market (DiMaggio et al., 1983). Buyers during the growth phase are so-called 'early adopters'.

The maturity phase of the PLC model is characterized by the technical maturity of the product. This is achieved through optimized production processes and a reduced rate of product innovation. The 'early majority' of the buyers accepts the basic concept. At this stage price and market structures are fairly stable. In addition, DiMaggio and Powell found a tendency towards homogeneity of organizational models after a market becomes well established. From several approaches in the initial stages of the life cycle, only a few dominant organizational models emerge (DiMaggio et al., 1983). In their eyes, this so-called institutional isomorphic change takes place due to three reasons: political influence, standard response to uncertainty, and professionalization.

The saturation phase can be detected by smaller increments in sales. Customers in this phase belong to the 'late majority'. A decrease in sales can be postponed by extension strategies (recruiting new customer groups and identifying new applications). Such strategies alter the shape of the life cycle curve and create follow-up cycles. Therefore, the length of this phase can vary widely.

The decline phase often starts with price cutting and repressive competition. This is quite similar to what was termed 'cutthroat competition' in the case of the technology shock. This signals that it is high time for suppliers to counter the decline in sales by further differentiation (Rifkin, 1994). Customers of this phase are called 'laggards'.

1.3.4 Conclusion - do technology shocks cause a PLC-like pattern?

In sections 1.3.1 through 1.3.3, it was found that technology shocks can indeed trigger phases that are similar to the PLC. While the PLC concept mainly stems from observations within the consumer market, researchers of technological change have mainly investigated industrial goods markets. Observables for the PLC are primarily sales, revenue, profit, gross margin, and, to some extent, firm counts and simple strategic considerations. Additional measurables like technological activities and investment behavior are analyzed during investigations of technological change.

Approaches for categorizing innovations are commonplace in the literature focusing on technological change. The literature dealing explicitly with the PLC indicates that the degree of 'newness' considerably influences the shape of the PLC. The definition of newness, however, is rather vague (Hoffmann, 1972; Mueller, 1997). The nature of innovation is not really differentiated (Tichy, 1991). The literature that deals in depth with technological change does exactly that. As mentioned in section 1.2, there are ample suggestions on how to categorize technological change: the innovation of product properties (Henderson et al., 1990), the change in value proposition (Christensen, 1997), or the effect on companies' competence (Tushman et al., 1986). Along these axes, technology shocks represent a high degree of technological change and as such come along with a high level of 'newness' as stated by Hoffmann (Hoffmann, 1972).

It is to be expected that a PLC will follow whenever a technology shock can be identified by the aid of the matrix described in this chapter[9].

1.4 Implications for research

In this chapter, 'technology shocks' have been introduced as a new terminology to integrate aspects of 'major innovations', 'technical discontinuities', '(radical) technological change', and 'disruptive technologies'. A framework consisting of a competence dimension and a technology dimension has been proposed based on previous concepts for characterizing technological change. The competence dimension expresses the effects of

[9] For managers of companies, this means that there is another tool to help evaluate the severity of technological change. In addition, by being able to generally anticipate the events in the competitive landscape, managers can now structure and refine their strategy based on developments in the companies' specific environment they detect.

new technologies on the core competence of a company. Changes in the design of products or changes in performance parameters express the technology dimension.

Taste influences the perception of performance for consumer goods. The perceived change in performance parameters becomes the dominating element. For investment goods, the perception of performance is less subject to taste (thus more objective) and corresponds to changes in the product design.

The framework integrates four of five aspects which have been addressed by previous approaches. The aspects covered are technological advance, competence, product properties, and value proposition through performance. The value chain aspect, an element of Afuah's 'hypercube' (1995), is not included.

In addition to these aspects, it is possible to apply the proposed framework to both consumer and investment goods markets because it contains two methods for to operationalizing technology. Technological change in investment goods markets can thus be measured from two perspectives and the observations on the degree of change can be cross-checked to avoid misinterpretation. This thesis makes use of this advantage by selecting the machine tool industry as a research object.

It has been found that technology shocks trigger product life cycles. Although the PLC was developed from a consumer goods perspective, it can be combined with the findings from research on technological change. Research on technology shocks suggests that wave patterns can be expected that follow a scheme of four phases, where competition especially in the third and fourth phase leads to a shakeout and a reduction of the number of firms. An increase in failures (e.g. takeovers, bankruptcies, and settlements) would indicate the final stages of the PLC. A time lag between the shock and a significant increase in the number of failures can be assumed.

2 Success factors for surviving technology shocks

2.1 Introduction

Several authors have suggested and investigated factors necessary for the successful management of technological change. Their perspectives range from an organizational focus ('attributes of the company') or a decision-oriented view ('strategic decisions') to a focus driven by change and implementation ('operational decisions'). Most of the findings are based on observations in fast-moving industries; some try to validate the statements beyond the computer industry (e.g. in the investment goods industry such as machine tools) or relate findings of company survival to technological change.

In the following subsections, the most prominent factors that seem to influence the successful management of technology shocks are extracted from past research. The studies are categorized according to whether they were carried out in the context of technological change, in the more general context of innovation, or other contexts. Each study is categorized according to whether it was conceptual or econometric in nature or if it was based on case studies. Those factors that were identified as significant in econometric studies on technological change form the basis for a model of successful management of technology shocks. This model will be tested quantitatively in chapter 6. Qualitative studies and quantitative research in similar contexts (e.g. product innovation) suggest that there are further variables that influence successful management of technology shocks. These factors are also taken into account in the following sections and will be used to expand the model in chapter 6.

2.2 Main theoretical and methodological streams

There are a number of theories concerning the concept of technology shocks. The neoclassical and organizational theories seem most applicable with regard to the economic and organizational aspects. Other theories include the decision theory (Laux, 1998) as well as community ecology (Astley, 1985). Most of the considerations for a successful mastering of the change in technology are based on the theories mentioned below.

The **neoclassical economic theory** differentiates between incremental and radical innovation. It suggests that in the case of radical technological change, entrants will replace incumbent firms since they will invest heavily in radical innovation. Their strategic incentive to do so is greater than that of established companies. The reason for this is that established companies have less incentive to actively promote[1] a new technology as they are enjoying a stream of profit from the old one. The issue of whether a monopolist is more likely to innovate than an entrant was the object of discussion between Reinganum and Gilbert/ Newbery (Gilbert et al., 1984; Gilbert et al., 1982). The 'patent race' model[2] they used consists of two firms, the monopolist and an entrant competing in R&D activities. The first firm that successfully innovates obtains a patent and exploits it. In the case of 'drastic innovation', where monopolists have no efficiency advantages, Reinganum suggests that there is a high chance that the entrants are the innovators replacing the established monopolists (Reinganum, 1983). On the other hand, in the case of a 'non-drastic innovation', the monopolist would be the innovator as his efficiency advantages dominate.

The **organizational theory** suggests that incumbent firms fail in the face of radical innovation because they are too inert and self-centered to provide the necessary prerequisites for survival and to react sufficiently to technology shocks (Burns et al., 1966). Their research efforts are less productive than those of the new entrants. Incumbents process information inappropriately as organizational change is risky and difficult to achieve (Nelson et al., 1982).

Henderson develops a model which integrates the neoclassical theory of investment behavior with organizational theory (Henderson, 1993). Elements of both neoclassical and classical organizational theories are reflected in the studies discussed in section 2.3 to ascertain success factors for surviving technology shocks (Brown et al., 1985; Cooper et al., 1992; Nadler et al., 1979).

Another approach to explain companies' relative success or failure is the **decision theory**. In this context, the process of decision-making and its motivations have been examined. Bourgeois and Eisenhardt summarize the two predominant views on how executive teams should make key decisions (Bourgeois et al., 1988). The 'rational-comprehensive' approach (Frederick-

1 Provided that the date of introduction of the new technology is influenced by the incumbents' investments.

2 For an introduction to patent races, see Chapter 10.2 'Introduction to Patent Races' (Tirole, 1988).

son et al., 1984) builds on a strong ex ante element - top management agrees on a framework of goal priorities. Alternatives are thoroughly researched in the preparation for a decision. Optimal choices on major issues are integrated into the existing strategy. The alternative approach is 'political incrementalism', which is a rather hands-on attitude of decision-makers to settle the issue temporarily (Simon, 1957) and find a solution over time (Quinn, 1980). An a priori goal consensus does not necessarily exist (Quinn, 1978); the search for decision alternatives is 'problemistic and constrained' (Cyert et al., 1963). Under this approach, strategy is the result of an adaptive step-by-step process, not really having an ultimate goal in mind. Lindblom termed this process 'the science of muddling through' (Lindblom, 1959).

The discussion in the previous chapter regarding how technology shocks initiate product life cycles and follow similar patterns as the PLC pointed at another organizational theory of relevance: **population ecology**. Parallels have been drawn between the life of biological organisms and the life of a product. A justification for the assumption of typical revenue curves for products was obtained based on the rules of growth observed in biological organisms (Hoffmann, 1972). The efforts to describe technology life cycles using tools of population ecology include the duration of the substitution process, industry demand, and market shares. Astley distinguishes between two ecological perspectives on organizational evolution - population ecology and **community ecology** (Astley, 1985).

Population ecology investigates evolutionary change within already existing populations. It describes survival as filtering out unfit members of the population and favoring only that subset of organizations optimally adapted to a given configuration of niche constraints (Hannan et al., 1977). DiMaggio and Powell tried to explain why organizational forms of companies assimilate over time. They termed this effect a 'mimetic process', a process which encourages organizations within a population to mimic each other as a way of dealing with uncertainty (DiMaggio et al., 1983). Population ecology emphasizes selective pressures that will eventually create a homogeneous set of organizational forms and a condition of population stability. By doing so, it does not, however, explain how populations come into being in the first place or how heterogeneous types of organizations are created. Community ecology tries to overcome these limitations by regarding populations as basic units of analysis within a larger community and focusing on the rise and fall of populations as basic units of evolutionary change. The survival of single populations within communities is determined by the degree to which a single population can function 'as a constituent member' of the community (Astley, 1985; Barnett, 1990).

Most organizational changes occur in the process in which new populations are born and old ones vanish. Technology shocks in terms of population ecology are 'discrete origins and extinctions'. They 'punctuate' longer periods of incremental or 'negligible change', called 'equilibrium'. The phenomenon of a technology shock has its equivalent in ecological evolution, 'which moves with an episodic, not gradual tempo. An abrupt branching of new lineages produces a stepwise pattern of change as opposed to a continuous one. Instead of replacing their ancestors through a steady process of transformation, new populations diverge to coexist alongside their ancestors until the latter are suddenly extinguished' (Astley, 1985). Carroll and Hannan found a correlation between the population density at the time of founding and the mortality rate within an industry (Hannan et al., 1977). This supports observations of the PLC that the number of companies actually seems to decline after reaching a peak.

Another effort to give product life cycles theoretical foundation is expressed in the **diffusion models**. Easingwood proposes nine different diffusion classes depending on the extent of external influence and a variable influence factor (Easingwood, 1988). This is an extension of the Bass diffusion model (Bass, 1969) and indicates how receptive different customer groups (late adopters, early adopters, etc.) are to innovation. His equation was supported by statistical tests on 20 data samples including the data on hybrid corn from Griliches' seminal article (Griliches, 1957). Diffusion models can explain the duration and intensity of technological change. The basic assumption of diffusion models in the context of technological change is that innovations are not implemented simultaneously throughout an industry. Diffusion processes happen for the distribution of information and the acceptance of new products (Easingwood, 1988). Information is costly and difficult to acquire because it is often highly specific and particular. Von Hippel referred to this phenomenon as 'sticky information' (von Hippel, 1994).

The propositions that underlie diffusion models for the adoption of innovation can be applied to the consumer as well (Gatignon et al., 1985). The knowledge about the existence and the capabilities of a new technology diffuse through an industry.

It was found that the rate of acceptance and substitution for subsequent generations of high-tech electronic products can be described using a diffusion model - new technologies diffuse through a population of potential buyers over time. The transfer of information about a new technology diffuses exponentially once the technology is introduced. The so-called 'bandwagon effect' describes a phenomenon in which the number of companies that accept a technology is a function of the number of companies already using that technology. Similar effects can be observed for customers (Norton et al., 1987). The result is the well-known S-curve for cumulated sales. In cases with relatively brief time intervals between successive technology generations, it also implies that the older technology continues to experience sales growth, increasing its market share after the new technology has already been introduced. In other words, when the old technology has not yet sufficiently penetrated the market, there will be a time lag between the outset of a technology shock and its ability to be detected as an influence on the market share[3] (Norton et al., 1987).

2.3 'When Darwin strikes': success factors for overcoming technology shocks

An overview of the literature reveals that some attributes for overcoming technology shocks must exist in companies (long) before a technology shock occurs. Not everything can be decided upon and arranged the moment a technology shock is identified.

With this in mind and accounting for the main theories supporting the observations in research, success factors can be summarized into three groups: attributes of the organization, strategic decisions[4], and operational decisions. Factors which have not yet been sufficiently addressed in the past research but have been proposed by managers will also be included in the research of this thesis.

The amount of literature referring directly to radical technological change is quite limited. For this reason, sources from a more general context of innovation as well as some very basic considerations or ones related to radical change in a broader framework are also taken into account. The type of study (qualitative, based on case study, econometric) is also indicated. The success factors as found in literature form the basis for subsequent hypotheses that are tested in this research. The three groups of success factors will be enriched by factors proposed by managers (chapter 3 and sections 5.2 and 5.3) to form a variable set that is the basis for creating a model to manage technology shocks in chapter 6.

---

3 Thus relying on market share to detect and to time technology shocks is misleading.

4 Christensen, Suarez and Utterback (1998) also used the category 'strategic variables' together with the categories 'conventional economic variables', 'population ecology variables', and 'technology variables'.

2.3.1 Company attributes prior to the shock

The ability to discover and exploit market opportunities - or innovation areas (Schrader, 1996) - is specific for any given person and organization as it depends on prior knowledge (Shane, forthcoming). Cooper and Smith advise corporations to develop some **sensitivity** to differences in **newly developing markets** (Cooper et al., 1992; Cooper, 1979). Their study - based on 27 cases of established companies in 8 industries[5] - concludes that established corporations are preoccupied due to their heavy investment in the older technology. Sunk costs keep them from advancing into the new field (sunk cost bias). The cases they present are quite similar in that the established companies are not able to prevent new technologies from taking over. In the study of Baker et al. (1986), the extent to which a project's business and technical goals were well-defined and widely recognized was not found to be significantly related to ultimate success or failure. Cooper and Kleinschmidt (1986), on the contrary, carried out a detailed investigation of the preparatory stage of successfully approaching innovation projects in their context of the 'NewProd' project, in which 123 firms participated. Based on their results, one can conclude that the intensity of market and technology planning does have an influence on the success of innovation.

The **ability to understand the status and trends of technologies** (Price, 1996) - necessary for achieving competitive advantage - also involves the ability to accurately assess the time and cost of converting market need into market demand (i.e. adequate understanding of technology push/ feasibility and market pull/ demand). A firm may need to develop 'absorptive capacity' in order to exploit extramural knowledge effectively (Cohen et al., 1990). As a side effect a company with absorptive capacity can thus also predict the nature of future technological advances more precisely. Unfortunately, absorptive capacity as a prerequisite for improved forecasts was not an object of Cohen and Levinthal's empirical analysis; testing this hypothesis is the object of future research.

Maintenance of the status quo is a scale that Ettlie (1983) constructed from the following three items - 'encouraging more formal meetings', 'sticking to

---

[5] Cooper and Smith looked at cases in the following industries: ball-point vs. fountain pens, CT scanners vs. X-ray nuclear medical equipment, diesel-electric vs. steam locomotives, electric vs. mechanical typewriters, electronic vs. electromechanical calculators, electronic vs. mechanical watches, microwave vs. gas/electric ovens, and transistors vs. receiving tubes. According to their survival or non-survival, market share information, and other secondary information on their business performance, companies were categorized as 'successful' (7 companies) or 'unsuccessful' (20 companies).

rules and protocols', and 'disciplining people who depart from routine'. As an expression of **resistance to innovation,** the scale is negatively correlated to the adoption of radical process innovation. Schumpeter (1912) already described resistance to innovation as something inherent to innovation. Witte (1973) describes this kind of resistance as a barrier of 'not knowing' and 'not wanting'. Resistance to innovation, for example, is expressed in the NIH (not-invented-here) syndrome (Katz et al., 1982).

In addition to maintaining the ability to forecast technology shocks, the **planning system** has to be flexible enough to incorporate sudden changes in the competitive and technological environments. Christensen calls this ability 'discovery-based planning' (Christensen, 1997). This is to some extent a deviation from the conventional point of view that planning processes should be systematic (Johne, 1984). However, Johne also acknowledged the necessity to include all company functions in the planning process.

As mentioned above in the explanation of architectural innovation, much of a company's knowledge has to do with its **information flow** - it is 'embedded in the companies` communication channels, information filters and problem-solving strategies' (Henderson et al., 1990). Sociological research on the success of 'Silicon Valley' and the relative failure of 'Route 128' companies emphasized the importance of an environment open to the permanent exchange of ideas and know-how between companies, particularly amongst non-executives. While companies on the East Coast in the US guarded their technical know-how very carefully and thus became insular organizations, Silicon Valley companies did not try to prevent their employees from exchanging ideas with their peers at other companies and helping each other solve technical questions (Saxanian, 1995). In a study on information trading in the mini mills and specialty steel industries, Schrader (1995) finds that technical managers tend to exchange technical information across firm boundaries (including competitors) when it appears beneficial to their own companies. Frequent and diverse communication which is not overly restricted by rules supports the success of technological innovation[6] (Allen, 1984; Ebadi et al., 1984) and is typical for companies in uncertain environments (Brown et al., 1985). Successfully managing technology shocks requires a corporate culture that promotes the enhancement of competence and the absorption of new capabilities (Cohen et al., 1994). There has to be a climate of continuous innovation in place for all matters regardless of their individual relative importance. The culture should focus on individual initiative and innovation (Nadler et al., 1979). Too many rules in the stages from idea

6 Further empirical results on the correlation of innovation success and information flow are found on pp. 360 in Hauschildt (1997).

generation to implementation can have a negative impact on success (Albers et al., 1991).

The priority[7] which the **management of technology** receives within an organization is critical when facing a technology shock. Management in firms that survived external crises typically **focused** more on external developments than companies that went bankrupt (D'Aveni et al., 1990). Price[8] even goes as far as to claim that the task of managing technology is at least integral to, if not synonymous with, strategic management. It 'links engineering, science, marketing, operations, human resources, and other management disciplines to formulate strategy, develop technological capabilities, and use them to achieve strategic objectives' (National-Research-Council, 1987; Price, 1996). As Price's claims are based only on two cases and a few brief examples, the validity of these statements still remains to be tested. The direct involvement of top management in the preparation of technology decisions seems to be critical. Albers and Eggers (1991) found that the influence of top management is strongly linked to success when innovation moves from realization to implementation. If top management gets involved too early, the results can turn out to be negative. In addition, Baker et al. (1986) confirmed the importance of involvement by top management. Chakrabarti (1974) claims the necessity of a 'champion' for successful product innovation, someone occupying a position within the company with sufficient knowledge about the innovation. This position together with knowledge regarding the market, technical competence, drive and aggressiveness, and political influence, defines the qualities of a 'champion'.

Cross-sectional studies of the relationship between firm size, market strength, and innovative activity in general found no systematic relationship (Baldwin et al., 1987; Cohen et al., 1989). Paolillo and Brown (1978) found no significant correlation between innovativeness and R&D budget size. Nevertheless, the size of a company expressed in annual sales and total assets as well as its **financial strength** (stock and bond ratings) is definitely a factor contributing to the firms' survival (Christensen et al., 1998; Smith et al.,

---

7 The decision-making process itself is independent of the Brim-Glass-Lavin-Goodman phase scheme as Witte (1968a), Witte (1968b) pointed out.

8 Price (1996) illustrates this in the example of US steelmaker NUCOR (according to an article by Richard Preston, "Annals of Enterprise, Hot Metal", The New Yorker, Feb. 25 & March 4, 1991) where the president of NUCOR, David Ayock, personally inspected the Schloemann-Semag pilot plant in Germany. The technical feasibility of continuous steel casting by its compact-strip-production steel casting machine was demonstrated at this location. The acceptance of this flexible and cost-reducing yet risky technology prepared the ground for NUCOR's current dominant position.

1988). In other words, a large company has some tolerance for mistakes before it disappears.

Following the diffusion process of semiconductor technology versus tube technology, Tilton (1971) suggests several factors which motivate new entrants: firstly, new companies have no vested interest in protecting old technology. Secondly, smaller independent organizations are highly motivated in capturing new and comparatively small markets, whereas new markets do not affect the overall growth and profit rate of large companies in a similar manner. Thirdly, established companies might be averse to risk. Developing new technology could mean reducing profits in their current operations. On the other hand, established companies do have an interest in defending their dominant positions by trying to dominate once again in the next generation technology. For this reason they maintain large R&D efforts within their main organizations.

The **ability to produce** and sell a product was part of a study by Baker et al. (1986). This ability has a positive effect on the possible success of innovation. Link (1987) collected managers' perceptions on the most important factors for the success and the failure of innovation. The two top-rated items were a new product showing synergy with existing marketing skills and one that showed synergy with existing technical and manufacturing skills[9]. In addition, involvement in the manufacture of a product contributes to the exploitation of new information relevant to that particular market. The ability to conduct complementary research in order to broaden a company's general background knowledge eventually helps to respond quickly to external changes. These are additional aspects which are addressed by the absorptive capacity of a company (Cohen et al., 1994).

Pfeffer and Salancik (1978) identified the **ability to acquire and maintain resources** to be key to organizational survival from a 'resource dependence perspective'.

An overview of these studies is provided in table 2.1. Under the header 'context' it is indicated whether the research setting is related to technological change (T.C.), the more general context of product innovation (P.I.), or other (O.). The 'type of study' describes whether the past research was conceptual (Cpt.), case study based, or econometric. The industries of the case studies and quantitative econometric studies are stated together with the size of the sample. '+' indicates a positive effect of the respective factor on company success, '-' a negative effect.

---

[9] This is further support for the 'competence' axis of the technology shock matrix introduced in chapter 1. The less synergy there is with existing skills, the more fundamental the impact of the new technology.

From the attributes listed in table 2.1, three will be included in a 'base model' in chapter 6. Absorptive capacity, resistance of innovation, and financial base are factors that have been tested in econometric studies and have shown significant impact on firm performance in the context of technological change. The other factors have not yet been tested statistically under these conditions and will be used as candidates to expand the base model.

Table 2.1: Past research on attributes of the company prior to the shock

		Context			Type of study			
Category and drivers	Contribution	T. C.	P. I.	O.	Cpt.	Case study based	Econometric, industry	Impact
Ability to forecast								
Ability to understand status and trends of technologies	Price (1996)	✓			✓	Böhm, Walmart and several examples		+
Absorptive capacity	Cohen and Levinthal (1990)	✓	✓				N = 318, manufacturing	+
Sensitivity to new markets	Cooper and Smith (1992)	✓				27 comps. in 8 industries		+
	Cooper (1979)		✓				N = 103, 'active in new prod. develpm.'	+
	Cooper and Kleinschmidt (1986)		✓				N = 123, various	+
Resistance to innovation								
Status quo maintenance	Ettlie (1983)	✓	✓				N =~40 food related	-
'Not-knowing' and 'not-wanting'	Witte (1973)		✓		✓			-
Flexibility of plan								
Ability of incorporating new discoveries into planning system	Christensen (1997)	✓			✓	Cases from disk drive and excavator inds.		+
Systematic generation of ideas and integration of several departments into planning process	Johne (1984)		✓				N = 16, electronic instruments	+

Table 2.1 (continuation)

		Context			Type of study			
Category and drivers	Contribution	T. C.	P. I.	O.	Cpt.	Case study based	Econometric, industry	Impact
Information flow								
Rigid communication channels and information filters	Hendersen and Clark (1990)	✓			✓	Photolitho-graphic ind.		-
Informal inter-company know-how exchange	Saxanian (1997)	✓			✓	Silicon Valley, Route 128		+
	Schrader (1995)			✓			N = 103, mini-mills, specialty steel	+
High degree of communication with external, and informal procedures	Brown and Utterback (1985)	✓				N = 6, food, paper, computer		+
	Christensen (1997)	✓			✓	Disk drive and excavator inds.		+
Frequency and diversity of communication, low degree of formalization	Ebadi and Utterback (1984)		✓			117 academic projects		+
Company culture								
Climate of continuous innovation and improvement, focus on individual initiative and innovation	Nadler and Tushman (1979)		✓		✓			+
	Cohen and Levinthal (1994)	✓	✓		✓			+
Extensive rules and methods for the implementation of ideas	Albers and Eggers (1991)		✓				N = 70, '8 innovative industries'	-
Management focus								
Technology management is equivalent to strategic management	National Research Council (1987)	✓			✓			+
	Price (1996)	✓			✓	Böhm, Walmart and several examples		+

Table 2.1 (continuation)

Category and drivers	Contribution	Context			Type of study			Impact
		T. C.	P. I.	O.	Cpt.	Case study based	Econometric, industry	
Involvement of top management	Albers and Eggers (1991)		✓				N = 70, '8 innovative industries'	-
	Baker, Green, and Bean (1985)		✓				N = 21, flat steel, chemicals, process food	+
Orientation towards external developments in times of crisis	D'Aveni and MacMillan			✓			N = 114, manufacturing, retailing, transport	+
Necessity of product champion	Chakrabarti (1974)		✓			45 cases at NASA		+
Financial base								
Financial strength (stock & bond ratings, net sales, total assets)	Smith and Cooper (1988)	✓					N = 29, microwaves, TV, calculator, transistor, CT scanner	+
	Christensen, Suárez, Utterback (1998)	✓				N = 51, disk drive ind.		+
Substantial efforts by incumbents to develop new technologies	Tilton (1971)	✓			✓	US semiconductor ind.		+
In-house manufacturing								
Experience in production	Baker, Green, and Bean (1986)		✓				N = 21, flat steel, chemicals, process food	+
Synergy with existing technical and productive skills	Link (1987)		✓				N = 135, largest Australian companies	+
Attractiveness for new hires								
Ability to acquire resources	Pfeffer and Salancik (1978)			✓	✓			+

T.C. = (radical) technological change or change and technology
P.I. = product innovation
O. = other
Cpt. = conceptual type of study
N = number of companies in sample

2.3.2 Strategic decisions in reaction to the shock

Managers of firms in the established, threatened industry must decide on how to respond to an innovation that has the potential to alter or destroy their companies' existing business. The **type of innovation**, whether in the product architecture as a whole or in a component of the design, influences chances for success. In the disk drive industry, for example, it is evident that firms that 'target new market segments with an architectural innovation will tend to be more successful than those that target existing markets or innovate in component technology, even controlling for all the competing predictors for survival' (Christensen et al., 1998). As mentioned in the previous chapter on frameworks categorizing technological change, the fact that a technological change is either knowledge-enhancing or knowledge-destroying does play a significant role in the success of a company (Tushman et al., 1986).

Christensen, Suárez and Utterback conclude that elements subject to managerial influence, rather than factors from the outside environment beyond the control of managers, were the primary factors driving firm survival in the disk drive industry. Besides incorporating the dominant design, the right **timing of entry** appears key to survival (Christensen, 1997; Smith et al., 1988). The 'window of opportunity' occurs towards the end of the growth stage, a concept described previously. Moving first is of advantage for companies that have already done a great deal of 'development work' (Shapiro et al., 1999). Klepper found a correlation between performance and early entry as a result of his research projects on factors governing the early evolution of industries and those shaping the evolution (Klepper et al., 1990). Klepper follows industries from their origin to the mature state. He divides the history of each industry into three phases: the first phase ends when the number of firms has reached its peak; the second phase when an approximate stability in number of firms has been attained. The second phase is characterized by a shake-out, but the findings (e.g. on firm age) are not explicitly linked to changes in technology. Other causes for shake-outs are included. These decisions concern the timing of entry, magnitude of commitment, degree of organizational separation, and competitive strategy for new businesses.

In the first phase, a wide range of factors are also presented that could lead to the beginning of an industry (state programs, change in customer taste and behavior, etc.). Reidenbach and Moak (1986) found that a longer period of preparation seemed positive for industrial success in the case of retail banks.

The history and experience of companies influence they way they respond to technology shocks (Cooper et al., 1992). The anecdotal evidence Cooper and Smith present in their case studies suggests that the **commitment of resources** in the new field, **separation** of evolvement from an established organization, and a **strategy** for the new business similar to the existing business strategy are important factors for success. These observations obviously require further investigation as they are based on anecdotal evidence. In addition, Christensen recommends that the **size of the organization** competing in the new market should match the size of the market (Christensen, 1997). In his study of success factors, Link (1987) found that managers expect the synergy of the new product with existing marketing skills to have the highest influence on the success of innovation as far as the **business model** is concerned.

Brown and Utterback reasoned that a high degree of uncertainty in the environment of firms, consisting of market, product, and process technologies, requires 'a greater degree of use of integrating roles[10] and departments for effective performance' (Brown et al., 1985). On the contrary, firms in environments of certainty can rely on formal hierarchies and procedures for success. While firms in a rapidly-changing environment are more **'organic' in structure**, firms in a more stable commercial and technical environment are more **'mechanistic'**. The organic structure enables a higher degree of interpersonal communication necessary during periods of high uncertainty (Brown et al., 1985). Based on the few successful cases where mainstream firms were able to establish a position in disruptive technologies, Christensen advocates **autonomous organizations**. He urges that responsibility be given to organizations whose customers appreciate disruptive technologies (Christensen, 1997). Established companies should allow the organization that deals with the disruption to remain independent (Bower et al., 1995). Innovations can be placed in quasi-entrepreneurial ventures outside of the parent company (Gaitanides et al., 1985).

Interfirm collaboration for the commercialization of complex goods has proved to be highly beneficial for the development and introduction/ marketing of new technologies. Mitchell and Singh investigated the hospital software systems industry and found that this holds true for incremental innovations. They discovered that when the environment changes gradually, firms using development-oriented and marketing-oriented collaborative relationships are less likely to shut down than businesses acting alone. However, when environmental shocks affect the content of collaboration, the chances of a company's survival are reduced. The study shows that businesses with collaborations directly affected by the shock had a higher probability of failure than companies not involved in such collaborations (Mitchell et al., 1996).

[10] They described the 'gatekeeper' as an example of such a role.

Appropriability (Teece et al., 1994) can take place through in-house development, licensing, cooperations, venture capital investments, and **acquisitions** (Corsten, 1982; Mittag, 1985).

An overview of past research on the strategic decisions described above is provided in table 2.2.

Table 2.2: Past research on strategic decisions

		Context			Type of Study			
Category and drivers	Contribution	T. C.	P. I.	O.	Cpt.	Case study based	Econometric, industry	Impact
Type of innovation								
Target new market with architectural innovation	Christensen, Suárez, Utterback (1998)	✓			✓		N = 51, disk drive ind.	+
Knowledge destroying change	Tushman and Anderson (1986)	✓					N = 72, cement, airlines, minicomputer	-
Timing of market entry								
Enter market during window of opportunity at the end of the growth stage	Christensen, Suárez, Utterback (1998)	✓			✓		N = 51, disk drive ind.	+
	Christensen (1997)	✓			✓	Cases from disk drive and excavator inds.		+
Enter market early	Klepper (1990)			✓				+
	Smith and Cooper (1988)	✓					N = 29, microwaves, TV, calculator, transistor, CT scanner	+
	Shapiro and Varian (1999)	✓				Cases from railroad, electricity, color TV		
Initiate technological change	Tushman and Anderson (1986)	✓					N = 72, cement, airlines, minicomputer	+
Longer and more intensive preparation	Reidenbach and Moak (1986)		✓				N = 121, retail banks	+

Table 2.2 (continuation)

		Context			Type of Study			
Category and drivers	Contribution	T. C.	P. I.	O.	Cpt.	Case study based	Econometric, industry	Impact
Organizational independence								
Flexible, non-hierarchical structure	Brown and Utterback (1985)	✓				N = 6, food, paper, computer		+
	Christensen (1997)	✓			✓	Cases from disk drive and excavator inds.		+
	Bower and Christensen (1995)	✓			✓	Examples from disk drive ind.		+
	Christensen (1997)	✓			✓	Cases from disk drive and excavator inds.		+
	Gaitanides and Wicher (1985)		✓		✓			+
Business model								
Competitive strategy for new market resembling strategy for established technology	Cooper and Smith (1992)	✓				27 comps. in 8 industries		+
Synergy with existing marketing skills	Link (1987)		✓				N = 135, largest Australian comps.	+
Acquisitions								
	Corsten (1982)		✓		✓			+
	Mittag (1985)		✓		✓			+
Alliances & Joint Ventures								
Interfirm collaboration	Saxanian (1995)	✓			✓	Silicon Valley, Route 128		+

T.C. = (radical) technological change or change and technology
P.I. = product innovation
O. = other
Cpt. = conceptual type of study
N = number of companies in sample

2.3.3 Operational decisions in reaction to the shock

Many scholars elaborate in general on the changes necessary within an organization after a technology shock has taken place. Tushman and Anderson conclude that additional research needs to deal with patterns of technological change mainly due to the limited sample size. Nonetheless, they found that discontinuous and incremental innovations need to be dealt with differently (Tushman et al., 1986). Incremental innovation reinforces the capabilities of established organizations, while radical innovation requires the creation new technical and commercial skills (Henderson et al., 1990; Tushman et al., 1986). Cooper and Smith expect major innovations to compel companies to develop **new resources and skills.** In the case of the CT scanner, these new skills that had to be acquired were computer technology and product design (Cooper et al., 1992). Pfeffer and Salancik propose a resource-dependent view - the ability to **acquire and maintain resources** is key to organizational survival (Pfeffer et al., 1978). Frameworks and routines become institutional devices in carrying out recurring tasks. Established companies have an interest in maintaining technical in-house capabilities to stay abreast with the competition (Tilton, 1971). Those companies with an emphasis on R&D are more likely to be successful (Smith et al., 1988). Since procedures and processes in established companies are fairly stable, there seems to be a need for **new problem-solving approaches** when a technology shock occurs (Henderson et al., 1990).

Cooperation with customers and suppliers[11] helps overcome technological change. In her study of service industries, De Brentani (1989) found that innovations need be consistent with customers' values. Von Hippel (1988) developed a concept to systematically exploit ideas of trendsetting customers pushing the development of a product through own modifications - the 'lead-user-concept'. The idea that customer feedback can enrich the value of the information is not a new one (Rubenstein et al., 1976; Witte, 1972)[12]. Baker et al. (1985) found that sources of ideas influence the success of development projects. If projects are solely initiated by R&D, there is much lower probability of success. If the initiative, however, comes from the customer or company functions closer to the customer, the chances of success increase. When companies exist in 'regional clusters' of competence, inter-firm collaboration automatically takes place through employee fluctuation and social contact (Saxanian, 1995).

Cohen and Levinthal (1990; 1994) observed that there is an increase in internal efforts (usually indicated by R&D spending) in situations where the 'ease of learning' regarding external development diminishes. The advanced technical **training** of employees and monitoring of technical literature contribute to the maintenance of absorptive capacity.

External R&D plays an important role in high-tech industries (Pisano et al., 1995). In particular, new entrants bring fresh know-how to an industry

---

[11] Harhoff (1991) shows (N = 132) that suppliers can also be an important source of innovation through knowledge spillovers in a study involving several industries.

[12] Hauschildt (1997), p. 202 f. provides an overview of further empirical studies that deal with innovation, cooperation and success.

which can be acquired through cooperations (Pisano, 1990). There are also views contrary to this. When a shock occurs, a company should have complete control over those areas that are 'shock affected' (Mitchell et al., 1996). Companies should develop as much know-how as possible internally. The process of managing a technology shock would require the organization skills to change, thus changing the organization itself.

Table 2.3 provides an overview on operational decisions.

Table 2.3: Past research on operational decisions

		Context			Type of Study			
Category and drivers	Contribution	T. C.	P. I.	O.	Cpt.	Case study based	Econometric, industry	Impact
Resource commitment								
New technical and commercial skills	Cooper and Smith (1992)	✓				27 comps. in 8 industries		+
	Henderson and Clark (1990)	✓			✓	Photolithographic ind.		
	Tushman and Anderson (1986)	✓					N = 72, cement, airlines, minicomputer	+
In-house technical capability to keep firms abreast	Tilton (1971)	✓			✓	US semiconductor ind.		+
R&D emphasis	Smith and Cooper (1988)	✓					N = 29, microwaves, TV, calculator, transistor, CT scanner	+
Match size of organization with size of market	Christensen (1997)	✓			✓	Cases from disk drive and excavator inds.		+
Level of hierarchy								
Flexible, non-hierarchical structure	Brown and Utterback (1985)	✓				N = 6, food, paper, computer		+
	Christensen (1997)	✓			✓	Cases from disk drive and excavator inds.		+
Extent of implementation program								
New problem-solving strategies	Henderson and Clark (1990)	✓			✓	Photolithographic ind.		+

Table 2.3 (continuation)

		Context			Type of Study			
Category and drivers	Contribution	T. C.	P. I.	O.	Cpt.	Case study based	Econometric, industry	Impact
Cooperation w. suppliers and customers								
Interfirm collaboration in regional clusters	Saxanian (1995)	✓			✓	Silicon Valley, Route 128		+
Complete control over marketing and development of shock affected areas	Mitchell and Singh (1996)	✓			✓		N = 973, US hospital software	+
Development sources are customers or functions that are close to the customer (e.g. sales)	Baker, Green, and Bean (1985)		✓				N = 21, flat steel, chemicals, process food	+
Customer orientation ('Consistence with customer values')	De Brentani (1989)		✓				N = 115, financial, management, logistic services	+
Exploit 'lead-user' know-how	Von Hippel (1988)		✓			Several high tech cases		+
Trainings								
Absorptive capacity	Cohen and Levinthal (1990)	✓	✓				N = 318, manufacturing	+
Technical learning	Cohen and Levinthal (1994)	✓	✓		✓			+
External R&D								
Acquiring know-how from new entrants through cooperations	Pisano (1990)	✓					N = 30, 96 projects in pharma	+
Complete control over marketing and development of shock-affected areas	Mitchell and Singh (1996)	✓			✓		N = 973, US hospital software	-
New hires								
Ability to acquire resources	Pfeffer and Salancik (1978)			✓	✓			+

T.C. = (radical) technological change or change and technology
P.I. = product innovation
O. = other
Cpt. = conceptual type of study
N = number of companies in sample

Four factors describing operational decisions have shown significant results in the statistical analysis of past research in the context of technological change. An increased commitment of skilled resources should have a positive impact on company success. Cooperation with suppliers and customers - especially with regard to control over marketing and development - will be part of the base model in chapter 6. Other aspects include training and the extent to which know-how was acquired through external R&D.

2.3.4 Factors with no significant impact on success

Christensen, Suárez and Utterback (1998) tested the significance of several success factors for survival in the disk drive industry. They found no strong evidence that the following environmental factors could be critical:

- The total number of firms active in the industry (as proposed by population ecology)
- Market size (a larger market feeds more firms - as proposed by ecologists)
- Industrial sales growth (faster-growing markets make it likely for a company to be attacked by the competition - as proposed by ecologists)

As already discussed in section 2.3.1, the test was carried out to distinguish between the two ends of the spectrum - a company either leaves the industry or stays. Thus, additional information is lost on how above factors correlate with a company's performance in managing a technology shock.

Another study by Smith and Cooper (Smith et al., 1988) based on the US microwave oven industry[13] found that the following factors do not influence the success of an organization when a technology shock occurs:

- The strategic importance to the parent company of an organization's business (rated by industrial executives on a 1-to-7 scale)
- Degree of threat - the percentage of a firm's existing revenue threatened by the young industry
- Extent to which the organization's business is related to other lines of business of the parent company (rated by industry executives on a 7-point scale)
- Company-initiated R&D expenditures as a percentage of net sales
- Business unit autonomy and functional authority measuring the degree of independence from the parent company (rated by industrial executives)

The authors suggest that these findings should be subject to further investigation because of the rather small sample size employed in the study as well as other reasons. In their article in 1992, Cooper and Smith recommended autonomy for businesses dealing with major new technologies (Cooper et al., 1992). Subsequent studies came up with contradictory results for business unit autonomy (Christensen, 1997). Moreover, in contrast to the findings above, Cohen and Levinthal argue that there is a correlation between R&D intensity and the absorptive capacity of companies (Cohen et al., 1990). As for strategy, Abernathy and Clark describe far-reaching effects of technology shocks on the elements of a company's strategy. A company's strategy ap-

---

[13] Microwave oven industry was studied in depth. The color television industry, pocket calculator industry, transistor industry and CT scanner industry were investigated in secondary sources. A of three high performers and three low performers were selected from each industry.

pears to change with the kind of innovation that occurs (Abernathy et al., 1985).

2.4 Implications for research

The variable categories 'attributes of the company prior to the shock', 'strategic decisions in reaction to the shock', and 'operational decisions in reaction to the shock' cover a wide range of managerial actions. The attributes evolve over an extended period of time in the company and are influenced by several past management decisions.

Nine factors have been selected to be included in a basic model for statistical tests. These factors have shown significant influence on firm performance in past econometric studies of technological change. Three of the nine factors are attributes: Absorptive capacity and a strong financial base had a positive effect on firm performance. Resistance to innovation influenced company performance adversely.

Two factors are strategic decisions: Targeting new markets with architectural innovation had a beneficial effect on firm performance, while knowledge-destroying changes had a negative effect on companies. The second factor is related to the timing of market entry. Entering the market during a window of opportunity at the end of the growth phase influences firm performance positively.

Four factors are from the group of operational decisions: An increasing extent of human resources committed to the new technology is expected to promote firm success. Cooperation with suppliers and customers should lead to success when it allows for control over marketing and development. Trainings and external R&D in cooperations are additional factors to include.

The other factors will be used to expand the base model in chapter 6. In the next step, qualitative interviews have to clarify whether or not there are issues important in modeling the successful management of technology shocks that have not yet been included in the list of factors from previous research. These qualitative interviews are carried out in the course of pre-tests.

3 Research gaps and methodology

3.1 Introduction

This chapter aims to identify further gaps in past research in addition to these already indicated in section 2.3.4 and describe the approach to data collection and analysis in this study.

Research gaps are identified by two methods: First, the context and scope of past studies are investigated to identify research limitations. Interviews with managers are also used to complete the list of relevant factors that should be included in the research. Findings regarding the shortcomings of past research are used to refine the approach of this study.

The second purpose of this chapter is to describe the goals, approach, and process of collecting data for this study. The description allows the reader to develop an idea of the basic conditions involved in this research. An assessment can then be made about the quality of data that is the basis for the analyses in chapters 3 through 6.

3.2 Need for additional research

3.2.1 Additional success factors for future empirical analysis

Several potential success factors[1] have not yet been included in any empirical research so far. There appears to be a lack in empirical studies that investigate the relevance of the respective factors in an environment of radical technological change among the sources referred to in chapter 2. One example is the importance of top management's direct involvement in the preparation of technology decisions. In the context of innovation, it seems to be one of the most critical factors to consider in future studies; it has not yet been examined in the context of radical technological change. The importance of top

---

[1] i.e. to the best knowledge of this author.

management's involvement on a regular basis, e.g. in mergers, acquisitions, and negotiations, is quoted by practitioners (Bane, 2000; Bunnel, 2000) and described in cases like the one on NUCOR (Price, 1996).

Bourgeois and Eisenhardt (Bourgeois et al., 1988) found three indicators for successful decision-making in the high-velocity environment of the microcomputer industry:

- Strategic decisions are made swiftly yet with care
- The CEOs and top management teams are simultaneously decisive and powerful
- Risk in decision-making is sought while implementation is safe and incremental

It is interesting to see the extent to which these findings hold true once they are taken out of the 'high-velocity environment' and applied to industrial environments with 'less velocity'.

In the description of the various phases of technology shocks provided in section 1.3, it became evident that attention tends to move from product to process technologies after a dominant design has been established. Based on conclusions drawn from the development in the US chip industry, Price (1996) advises that management needs to focus its attention equally on product and process technologies (in particular on information technologies). The observation Price made applies to a stage where dominant players have already established themselves. So the question remains: 'To what extent is this advice still justified in the early stages of a technology shock during introduction and 'rapid growth'? Or may the opposite be more advisable, i.e. concentration on product technologies during these stages?'

Price emphasized the importance of the value chain, dubbing it the 'technological food chain' (Price, 1996). Two aspects are relevant for investigation: First, the extent to which it is justified to limit research to one product category - as in most previous research - thus running the risk of not detecting a shift in the value chain. For example, a supplier takes over a business segment from his customer. Also important is the degree to which the quality of relationships between the players in the value chain influences a company's success in surviving a technology shock. Abernathy and Clark also included supplier relations in their considerations (Abernathy et al., 1985).

The degree of diversification among incumbent firms as a factor for success in the face of an emerging young industry was suggested by Smith and Cooper (Smith et al., 1988), however, their study was lacking strong evidence. It would, therefore, be useful to test this indicator once again. Moreover, the recent success of Cisco diversifying in new fields of technology through acquisition has posed new questions on previous research (Bunnel, 2000). The opinion so far was that diversification through acquisition as opposed to in-house development would not lead to success.

Another aspect not yet extensively examined is the practical importance of convergence[2] effects such as that of telecommunications, information and computer technology and television networks that are currently taking place. Most research has instead focused on industrial segments that can easily be isolated. Converging industries could have the potential of yielding additional insight[3] into the management of technological change.

3.2.2 Designs of existing research and their limitations

Many of the analyses are exploratory or qualitative in effect - their sampling is limited and leads in part to contradictory results. Further research including cases of relative success and industries in environments of lower velocity seems to be necessary. Only a few studies have been able to gain sufficient data from a single industry (Christensen et al., 1998; Mitchell et al., 1996). When investigating radical technological change, it seems to be necessary to compare the cause of the shock. Ideally, there should be enough data collected from companies affected by the same shock. Much of the research quoted above suffers from limited data availability.

Apart from the nine variables that are considered for the base model (see section 2.4), there are several factors that require additional testing in the context of radical technological change. Most of the quantitative research used the number of companies combined with exits and entries as an endogenous observable (Carroll et al., 1989; Christensen et al., 1996; Christensen et al., 1998; Mitchell, 1989; Willard et al., 1985).

A large portion of the research is based on case studies (Abernathy et al., 1985; Cooper et al., 1976; Leonard-Barton, 1992; Levitt, 1960; Myers et al., 1978; Price, 1996; Quinn, 1978; Tilton, 1971). Case study design as a method of research is well-suited to exploring, describing, identifying and testing causal relationships (Yin, 1984). A design which offers better quality information is the 'embedded multiple case' as termed by Bourgeois and Eisenhardt[4] (Bourgeois et al., 1988)[5].

---

2 This effect was indeed discovered in the machine tool industry in the course of this research. See figure 4.6.

3 Convergence has also been observed in the machine tool industry that is in the focus of analysis here (see section 4.5.2).

4 Four cases in the microcomputer industry to study strategic decision processes

5 Embedded design denotes several units of analysis - the investigation is conducted at different levels. In this case, the levels consist of: the firm (or division) and its attributes, the firms' environment, the decision-makers, the decisions themselves, and their consequences. Embedded designs are complex, but they allow observations to be explained in greater detail and from various perspectives. Multiple case

There appears to be an additional need for research to investigate the first two stages of technology shocks. The phenomena surrounding dominant design have been subjected to ample research carried out by Utterback and others (Abernathy et al., 1978; Anderson et al., 1990; Christensen et al., 1998; Lee et al., 1995; Suarez et al., 1995; Utterback, 1994). Even more attention has been given to the dynamics of competition in existing markets (Porter, 1980). Thus, future investigations should also focus on the 'introduction' and 'growth' phases in an attempt to discover a way in which established companies can successfully enter into a new market.

3.2.3 Investigating complex industries

Industries that were studied in view of technological changes either possessed good documentation with ample information on companies and product histories or had short reinvestment cycles. In the latter case, the effects of innovation are easier to detect. There relevant events can be observed in a comparatively short period of time. From the table below, it is clear that of the studies related to technological change, only the study done by Cohen and Levinthal (1990) investigates an industry with a comparatively long reinvestment cycle. The other seven studies cover either mixtures of industries or rather short-lived industries, such as the electronics, television or disk drive industries. In the same way, quantitative studies that deal with innovation in general are based on mixed samples or industries where innovations establish themselves rapidly. Table 3.1 states the context of the studies (technological change, product innovation, or other), the number of companies, and the industries included in the sample. 'Type' indicates own estimates about the length of the reinvestment cycles[6] of the industries. It differentiates between rather long ('l') and rather short ('s') cycles. 'm' is used when the industries in the sample represent a mix of durations.

designs allow each case study to confirm or dispute the inferences drawn from previous ones. Multiple cases do not provide the basis for statistical treatment, but they permit more reliable models to be inducted than single case studies.

[6] The length of the reinvestment cycle is taken as an indicator for the speed of development and change in an industry.

Table 3.1: Types of industries investigated in quantitative studies

Contribution	Context			Number of companies, industry	Type
	T. C.	P. I.	O.		
Cohen and Levinthal (1990)	✓	✓		N = 318, manufacturing	l
Cooper (1979)		✓		N = 103, 'active in new product development'	m
Cooper and Kleinschmidt (198(		✓		N = 123, various	m
Ettlie (1983)	✓	✓		N =~40, food-related	m
Johne (1984)		✓		N = 16, electronic instruments	s
Schrader (1995)			✓	N = 103, mini mills, specialty steel	m
Ebadi and Utterback (1984)		✓		117 academic projects	m
Albers and Eggers (1991)		✓		N = 70, '8 innovative industries'	m
Baker, Green, and Bean (1985)		✓		N = 21, flat steel, chemicals, processed food	m
D'Aveni and MacMillan (1990			✓	N = 114, manufacturing, retail, transport	m
Smith and Cooper (1988)	✓			N = 29, microwaves, television , calculator, transistor, CT scanner	s
Link (1987)		✓		N = 135, largest Australian companies	m
Christensen, Suárez, Utterback (1998)	✓			N = 51, entire disk drive industry	s
Tushman and Anderson (1986)	✓			N = 72, cement, airlines, mini-computer	m
Reidenbach and Moak (1986)		✓		N = 121, retail banks	s
Mitchell and Singh (1996)	✓			N = 973, US hospital software	s
De Brentani (1989)		✓		N = 115, financial, management, logistic services	m
Pisano (1990)	✓			N = 30, 96 pharmaceutical projects	s

T.C. = (radical) technological change or change and technology
P.I. = product innovation
O. = other
l = industry with a comparatively long reinvestment cycle, s = industry with a comparatively short reinvestment cycle, m = companies from a mix of different industries

There seems to be a need to collect single industry samples that are large enough to do quantitative analysis on the same technology shock. Naturally, observations of repeated technological change in industries experiencing 'long waves' pose additional difficulties in collecting information i.e. data not being archived electronically and thus not easily found, firm records and archives having been deleted, witnesses of the events no longer being alive, etc. It is much more difficult to analyze events of decades ago than those of the recent past. In addition, the perception of cause and effect is far simpler for interviewees who had experienced both while they were still in business. It is very difficult to relate cause and effect in instances where the managers that had been involved have left their positions before the effects became visible. A similar situation was found in the machine tool industry[7]. Only a

[7] Similar to the telecommunications infrastructure industry.

few executives had been in a position of experiencing the comprehensive development of the industries in their recent past. It seems that a pattern of development is easier to trace in short-cycle or 'fast-moving' industries.

3.3 Goals and approach of the research process

Empirical evidence was collected to satisfy three goals. First, the machine tool industry as the object of research has to be described sufficiently. Data then has to be supplied as the basis for the statistical analysis of the factors that determine the performance of companies after technology shocks. Third, managers should comment on future technological trends to get an impression of the practical relevance of the topic[8]. Different approaches were used to obtain the data depending on the type of goal.

First, qualitative information regarding the research object[9] - the machine tool industry - was collected in order to describe the influence of technological change on the market structure, identify radical technological change in the recent history of the industry, and validate the categories describing technology shocks. Around 100 interviews within the machine tool industry[10] were carried out to collect and validate the information. The names of the companies are listed in the appendix of the corresponding chapter 4 'The machine tool industry and the effects of technological change'. Analyzing the technological and competitive background of the industry's history helped to time the shock events and to define periods of cross-sectional analysis. Extensive desk research supplemented the qualitative information. The need for such research is illustrated in the following - the interviews offered additional information on takeovers, bankruptcy, and settlements with suppliers, etc. As information regarding the events was sometimes inconsistent, it was necessary to cross-check it several times. In other words, a company's history had to be verified repeatedly.

In the next step, decision-makers of the machine tool industry's most significant companies were approached and asked to complete a detailed ques-

---

[8] The auspices of potential future shocks illustrate the applicability of the research results.

[9] The same process was carried out in the telecommunications infrastructure industry. It is assumed that the logic underlying this research on technology shocks in the machine tool industry can be applied to other industries as well. The findings from the research in the telecommunications infrastructure industry are not the focus of this work, though. Including them would reduce readability.

[10] About 20 interviews were carried out in other industries - mainly in the telecommunications infrastructure industry.

tionnaire on the management of technological change in their companies. This second goal was the most significant in the project. The relevance of different success factors in the management of radical technological change was to be identified. The factors themselves were extracted from previous academic research or observations made by practitioners and industrial experts as described in the earlier chapters. As described in the first two chapters, previous analyses generally focused on individually-selected success factors. In empirical research, the drawback has often been the limited availability of secondary data and the difficulty of generating additional primary data. Moreover, research frequently aims at testing theoretical models that focus on only one specific factor. So far, efforts have not been made to identify the most important factors from among a wide range of variables. Asking high level decision-makers to complete a 10-page list of questions was a challenge in itself (Dillman, 1978)[11]. After several 'motivational' contacts over the phone or in person, 63 decision-makers from 58 companies were willing to participate and deliver complete datasets for the investigation[12]. The companies were viewed as cohorts[13] - information about the whole group was collected for the total observation period. Thus, data was supplied by one and the same company before and after having experienced the technical shocks. In this sense, the setup differs from approaches in which the total number of companies in an industry is taken as a measurable. The companies that compose an industry are changing. Companies enter and leave the market. In the current approach, the goal has been to see how a group of companies develops over time under the influence of three technology shocks.

The replies to the structured part of the questionnaires serve as a basis for the descriptive and multivariate analytical statistics found in the later chapters. The companies that were approached and those that eventually participated are listed in table A.5.7 in the appendix. Decision-makers who participated in this study were reassured that their information would not be linked to their names[14]. No additional information about the participants will be published here.

---

[11] As incentives, the executives were promised (and sent) the results of the study. Some executives told about other studies that enclosed a ten dollar bill as a 'symbolic' reward. These executives had liked the idea. On the contrary, study results were expected anyway. This did not provide motivation enough for contacts to participate that were skeptical about the study from the outset.

[12] Questionnaires from only two companies could be collected from the telecommunications infrastructure industry.

[13] Cook and Campbell (1979).

[14] According to the ICC/ESOMAR (European Society for Opinion and Marketing Research) international code of marketing and social research practice, anonymity

Thirdly, the practical relevance of the questions approached scientifically throughout this thesis was tested by means of a Delphi study. Decision-makers in the machine tool industry were asked to voice their opinions about future technological change. Some of them also took part in the questionnaire-based survey and expressed their opinions in the open answer sections. Others were only interested in participating in the Delphi study. The results of the Delphi study are summarized in section 4.8.

3.4 The pre-tests

The process of compiling suitable questionnaires consisted of several rounds of pre-study. The potential success factors had been extracted from literature presented in the previous chapter. A first round of semi-structured interviews was carried out with 12 industrial experts and executives from the machine tool industry[15] (consulting companies like UBM and Mercer, and machine tool manufacturers such as Schaudt and Burckhardt&Weber).

To avoid having a questionnaire that could only be applied in the machine tool industry, an attempt was made to create a list of questions that would be as independent as possible of any particular industry. In this context, interviews and pre-tests were also carried out with 16 executives in various functions in the telecommunications infrastructure industry and related sectors[16] (Alcatel SEL, Dittberner&Associates, Enron Broadband Systems, GE Capital, Lucent Technologies, Nortel Networks, and Siemens AG). This first round showed that success factors as they appeared in literature were too generic, theoretical, and not specific enough to be applied directly in the form of a question such as 'To which extent does this factor apply to your com-

---

can be expected by the participants, see http://www.esomar.nl/guidelines/ICC_ESOMARcode.htm.

15 Many thanks to the quick and thorough answers received from Dzems Bruvelis (UBM), Alfred Dax (Societät für Unternehmensführung), Dr. Heinz Drink, Dr. Hans Klein, Klaus Leinemann, Ulrich Vetter, Jens Alber (like all the subsequent - Mercer Management Consulting), Dr. Jan Dannenberg, Oliver Herweg, Ralf Kalmbach, Thomas Kautzsch, and Wolfgang Stöckert.

16 I would especially like to thank Dr.-Ing. Wolfgang Peters (Alcatel SEL), Mr. Karl Dittberner (Dittberner&Associates, a Washington D.C. based market research firm publishing detailed annual industry surveys), John Brindle, Roger Strada, Russell Woody (all Enron Broadband Services), Prof. Dr. Tom Peisl (GE Capital, now Munich University of Applied Sciences), Bill Troutman, Hung Ling (both Lucent), John Strimas, Mark Mishii (both Nortel), Dr. Gerhard Baur, Mr. Hans-Eugen Binder, Mr. Werner Hutter, Monika Mahnke, Dr. Klaus Schmid, Mr. Nik Skaperta (all Siemens AG).

pany?'. Moreover, terms like 'absorptive capacity' (Cohen et al., 1990) were often not understood without detailed explanations. Finding qualified decision-makers who could respond to the entire range of questions was very time-consuming. Despite the difficulty involved in meeting decision-makers due to their time restraints, most reserved a few hours to one day for a personal interview for this study. It seemed that after having overcome the initial barrier, decision-makers tended to get highly involved in the subject and were glad to illustrate their answers with ample anecdotal evidence.

Practitioner suggestions relating to the degree of detail asked for in the study were taken into account. For example, it turned out that it would be beneficial to differentiate between subordinate and main/ strategic planning systems in a company. There may be situations in which strategic plans are changed only twice a year - even during periods of radical technological change - but subordinate plans are altered more frequently.

The practitioners also strongly recommended that the following factors be added to the research:

- The distribution of power in the company: It is expected that company performance will depend on whether or not top management considerations are implemented or if suggestions from lower levels within the organization are realized. In many cases, the decisions of top management are altered - 'improved' or streamlined by lower ranks in the hierarchy (based on their more hands-on experience).
- The consequences of decision-making: The decision-making process in many companies is not linear in the sense that decisions go through repeated loops of discussions with different participants, resulting in delays and mistakes in execution.
- The extent of decisions: Different results in company performance are expected depending on whether decisions are related to incremental or far-reaching changes in the company[17].
- The first goal of subsequent rounds was aimed at creating questions specific enough for practitioners to reply to. The next objective was to standardize the questionnaire in such a way that the interviewees could complete it without much instruction. The aim was to reduce the time required to go through the questions from one day to approximately 2 hours total for both a 'motivational' and a 'verifying' telephone call (Pohlmann, 1992). The time needed to establish contacts for interviews either in person or over the phone remained the same. Contacts were primarily established through recommendation.

[17] This differentiation is in line with the theoretical differentiation in decision theory: 'political incrementalism' versus 'the rational-comprehensive approach'

Observations in the subsequent pre-tests dealt with the range of and response to various rating scales, the layout and wording of questions, the terminology used to describe the beginning and end of a scale, etc.

The questionnaire reflects a two-fold approach - each item asks for a qualitative description of a success factor as well as a rating regarding its importance in view of the relative success/ failure of the respective company.

The scale that was preferred consisted of 5 items, ranging from 1 to 5. In the pre-tests, scales of 6 or 7 items were also tested. However, this broader range was preferred by only one out of 22 participants. In addition, the higher level of precision tended to confuse the interviewees. The scale would need to start with 1 and consist of an uneven number of items so that participants would read the descriptions at the beginning and end of the scale and not confuse '0' with a 'neutral' reply. In this way, it would also be possible to indicate a 'medium' rating. An even number of items would make it impossible for participants to give a 'medium rating'. The interviewees perceived this as a shortcoming.

In questions regarding the importance of the various factors, differentiating the ends of the scale as 'high' and 'low' worked out fine. Participants preferred to have a scale of the same length for all questions that required ratings. If questions qualified different factors, the ends of the scale had to be assigned specific terms on an individual basis. Some interviewees felt confused when the ends of the scale did not correlate with their intuitive response. An example of this was the question 'To which extent did resistance to change take place in your company?' In this case, participants would have intuitively placed 'low resistance' at the left end (value 1) and 'high resistance' at the right end (value 5) of the scale. Thus, a favorable outcome does not always appear on the right end and the less favorable one on the left.

Whenever the relevance of an individual factor was asked for in a neutral manner as in, 'Relevance of ... for the relative success/ failure of the company?', the rating was often simply forgotten. It proved more useful to address the interviewee directly, i.e. 'How do you rate the relevance of ... for the relative success/ failure of your company?'

In addition, the use of highlights reduced the reading time for the questions.

3.5 Making the questionnaire available

The questionnaire[18] was made available to participants in a variety of forms. Management had the choice of submitting three types of responses - online/ web-based, e-mail; and paper-based (standard mail, fax). The most efficient method of collecting data was intended to be via online/ web-based questionnaires (Stewart, 1999). An on-line questionnaire has the potential to reduce time and cost of the data collection and offers the possibility to respond simplified and standardized (Coomber, 1997; Sheehan et al., 1999). In order to avoid common media-specific errors (Kulp et al., 1999), pre-testing of the online version had to be particularly thorough. The standard procedure of duplicating electronic questionnaires in HTML was carried out to collect data[19] (Theobald et al., 1998). The completed questionnaires were returned via e-mail and, after a 'manual' cross-check[20], were imported into the database created automatically by the same software that had created the HTML questionnaires. Apart from technical differences, the electronic versions were treated as alternatives to mail and fax for the collection of information wherever possible (Bachmann et al., 1996; Mueller et al., 1996). After a 'motivational call' during which decision-makers were informed about the goal, background, and process of the study, the participants were able to choose the mode of receiving the questionnaire - either as a printout by mail or fax, or as a Word® document in the form of an e-mail attachment. A cover letter indicated the possibility of filling out the questionnaire online and provided the appropriate Internet link. Moreover, the cover letters were personalized and made reference to the introductory - 'motivational' - telephone call that had taken place. The online HTML questionnaire contained the same questions as the Word® version but was interactive. It could be accessed via the institute's homepage. A screenshot of the homepage with links to the questionnaires in English and German is included in the appendix of this chapter. Screenshots of the first pages of the English online HTML questionnaire as

[18] The questions and the data format expected in the answers as well as the possible answer ranges are displayed in appendix A.5.1 through A.5.3. The complete content of the questionnaire is contained in these three tables. The introductory statements to the questionnaire can be seen on screenshot A.3.2: The layout of the questionnaire can be seen at http://www.inno-tec.bwl.uni-muenchen.de/mitarbeiter /arnold/Machine%20tools%203A.htm.

[19] The basic version was created by Perseus Survey Solutions® (product description at http://www.perseus.com), which was generously provided by Mercer Management Consulting, Lexington/ Boston office. The HTML version was modified and put on-line by Macromedia Dreamweaver® provided by INNOtec.

[20] Also to avoid duplicates as described in Steele, Schwending and Kilpatrick (1992).

well as other technical details to be considered can also be found there. Paper-based responses were manually entered into the same database in which the electronic answers were stored. Several cross-checks (Erdos, 1993) were carried out to make sure that the data had been transferred correctly.

3.6 'Top management still not online'

Despite a wide choice of available response options, a number of participants still refused to put anything in writing. They preferred to relate the information over the phone or during a personal meeting. Table 3.2 summarizes the options that were used to submit responses to the questionnaires.

Table 3.2: Answer modes for the completed questionnaires

Format received Format sent	Online	E-mail	Fax	Mail	Phone	In person	Total
E-mail	2 (o:2)	19 (e:19)	11 (e:4; m:7)	8 (m:8)	1 (t:1)		41
Fax			5 (e:1; m:4)	1 (pers:1)			6
Mail			3 (m:3)	7 (m:7)	1 (t:1)		11
Telephone					1 (t:1)	1 (pers:1)	2
In person			1 (m:1)			2 (pers:2)	3

In parentheses: mode of filling out the questionnaire - o-online, e-electronically, m-manually, t-by telephone, pers-in person.

Out of a total of 63 replies, retired executives filled out 11 questionnaires while the other participants are still active in their jobs. Nine of the retired executives sent their questionnaires by standard mail, one by fax, and one by e-mail[21]. Even without taking the retired executives into account, 40% of the decision-makers (21 out of the 52 active executives) used only electronic media (e-mail or online web-based questionnaire) to respond. Only two executives made use of the web-based possibility despite the fact that they had

[21] Of this group, the one participant who had received the questionnaire by fax represented a very interesting case. He transferred the questions to his PC, printed out the answers, and returned the questionnaire by fax.

received a direct link to the on-line questionnaire[22] via e-mail - which they only needed to click on - as well as having the cover letter explaining the on-line HTML option. In addition, it had been explained to them that this HTML version was the fastest way of replying; however, and against expectation (Kanuk et al., 1995), this information was not much appreciated.

3.7 The necessity of networking

As expected, collecting questionnaires proved to be extremely difficult[23]. Although the subject in general appeared to be of considerable interest to most of the interviewees, the range of the questions together with the fact that a large part was considered by many to be 'confidential information' made the data collection process quite time-consuming. The overview of the feedback and response rates for the machine tool industry is illustrated in table 3.3.

Without the assistance of a number of industrial experts and managers who supported the study and established contact with several decision-makers, the necessary critical level of questionnaires would not have been reached. Executives contacted through recommendation filled out 40 of the 64 questionnaires[24] whereas 69% of contacts without personal reference in machine tool companies refused to respond. Only 15% were willing to fill out questionnaires, the rest offered to be of some assistance but declined to complete the questionnaires. The general willingness to assist was higher among institutions. Only 21% of the contacts resulted in no support or information.

Establishing contact with decision-makers through reference also beat the alternative of personal contact at industry and trade shows. The latter proved to be only slightly more successful than approaching decision-makers without previous contact.

---

[22] http://www.inno-tec.bwl.uni-muenchen.de/mitarbeiter/arnold/index.html.

[23] It took nearly a year and a half to collect data from decision-makers and industrial experts.

[24] I am especially grateful for the support of Peter Baumgartner (Schneeberger AG, CH), Dzems Bruvelis (UBM, D), Dr. Heinz Drink, Andre Heutschi (Schneeberger AG, CH), Mr. Werner Hutter, Jean-Pierre Jost (Schneeberger AG, CH), Richard Kegg, Rolf Kölblin, Prof. Dr. Hans-Ulrich Küpper, Klaus Leinemann, Dr. Eugene Merchant, Dr. Mushardt (Schleifring, D), Dr. Schäfer, Dr. Schimke (Kienbaum, D), Erhard Schmitt-Staubach, Dr. Treib (Agie Charmilles, CH), Ulrich Vetter, and Dr. Udo Weigel (Samputensili, I).

Table 3.3: Feedback and response rates for the machine tool industry

	Total number of contacts	Questionnaires filled out and information given	Willingness to support or information given	No info or support	Contact not pursued
Machine tool manufacturers	276	62 (22%)	48 (17%)	159 (58%)	7 (3%)
with referral	133	39 (29%)	26 (20%)	61 (46%)	7 (5%)
with trade show (EMO) contact	37	7 (19%)	5 (14%)	25 (68%)	
without referral or t.s. contact	106	16 (15%)	17 (16%)	73 (69%)	
Institutions & independent experts	19	1 (5%)	14 (64%)	4 (21%)	
with referral	12	1 (8%)	9 (75%)	2 (16%)	
without referral	7	0	5 (71%)	2 (29%)	
Total	296	64 (22%)	62 (21%)	163 (55%)	7 (2%)

3.8 'Going beyond the fruit fly': implications for research

Fruit flies[25] have been important test animals in biological research. They are chosen because of their rapid change of generations, the limited complexity of their organisms, and their availability as test animals. For example, observations about heredity transmission can be made in a comparatively short time.

Most of the past research on technological change has focused on industries with a high velocity of change (Bourgeois et al., 1988) and rather homogeneous structures (Christensen, 1997; Christensen et al., 1998). In a sense, these industries have become the 'fruit flies' for this kind of research. There is a need to study technological change in less specific environments. Therefore, one of the objectives for this research was to test the importance

[25] The fruit fly has the Latin name Drosophila. About its importance and limitations for biological research see http://www.drosophilahistory.com/sturtevant-dobzhansky-novitski/frame.html.

of success factors for the survival of technology shocks in an industry which had received little attention (quantitatively and empirically) thus far in this respect.

Previous studies have tried to avoid the problem of limited data sets by including data from a variety of different industries. Despite great difficulties, the aim of this research was to generate as much data as possible from within the same industry to ensure that companies were affected by the same type and degree of radical change. In addition, further potential success factors had to be considered that were brought forward by practitioners or evident in company histories. These have not yet been analyzed empirically, however.

Opening up several channels of information exchange with respondents from top management did not contribute greatly to the information collected. With the help of personal references, though, the sample size reached a level at which statistical analysis became feasible.

Practitioners have stressed the importance of taking factors into account that have not played a part in previous research. Thus the 'distribution of power', 'consequence of decision-making', and 'extent of decisions' are also included in the set of variables that will be tested.

4 The machine tool industry and the effects of technological change

4.1 Introduction

Rather long reinvestment cycles of about 15 years have created the notion that innovation in the machine tool industry is of incremental nature. But looking at its recent history, the integration of digital controls technology and computers into machine tools have hit the industry in three waves of shocks since 1955. Most companies underestimated the impact of the new technology. This chapter summarizes the post-war history of the machine tool industry since the invention and introduction of numerical controls and analyzes the disruptive character of this new technology on the market. About 100 interviews were conducted with decision-makers and industry experts who witnessed the development of the industry over the last forty years. A connection between radical technological change, industry structure, and competitive environment is established in this chapter. A number of important occurrences and interrelations are revealed that were previously overlooked in other studies. Qualitative findings from the questionnaire survey are included here. A historical perspective is used in this chapter to describe the machine tool industry as the object of this research In addition, it attempts to apply the technology shock framework established in chapter 1 to shocks that happened during the history of this industry. Section 4.8, which is based on a Delphi study (Arnold, 2002), takes on a future oriented perspective. It shows that technological changes that potentially have the effect of a shock can be expected for the machine tool industry in the future as well. This prognosis section emphasizes the relevance of the research for practical use.

4.2 The machine tool industry after World War II

While contributing to only about 2% of the national industrial production of the top ten machine tool producing countries[1] (Rendeiro, 1985; VDW, 2002), the machine tool industry is fundamental for the investment goods industry. It provides the principal industrial equipment base for manufacturing. The machine tool industry developed and matured along with its main markets, namely the machinery industry, the automotive industry, the defense and aerospace industry and other investment goods industries (BCG, 1990).

By the turn of the twentieth century, the United States had developed a strong position in the world's export markets. The World Wars strengthened the position of US manufacturers even further. Germany, which had been the world's largest exporter in 1910 (Mazzoleni, 1998), dropped behind. After World War II, much of the industry in Germany had been dismantled[2]. The United Kingdom, which had led the development of machine tools since the industrial revolution[3], was in the fortunate position to have had a manufacturing infrastructure that was fairly untouched by the war.

The destruction of the rest of Europe's manufacturing industries made the availability of equipment a fundamental part of manufacturing companies' post-war strategies[4]. Equipment was badly needed at almost any price. This allowed the industry to grow at a steady pace of approximately 10% annually from 1950 - 1970.

Technological change was incremental in character until the 1970s when the appearance of numerical controls upset the industry and caused considerable problems for established companies. The emergence of numerical control caused a major discontinuity in the machine tool industry's traditional product lines. What Christensen had observed for the rigid disk drive industry (Christensen, 1993)[5] happened in the machine tool industry in the face of radical technological change. Many of the then established compa-

---

1 Average per year as a percentage of total industrial production between 1955 and 2000.

2 For the effect of World War II and the subsequent 'dismantling' of most of Germany's production capacities see pp. 465 - 488 in Spur (1991).

3 The history of machine tools and the machine tool industry from 1700 to 1910 is described in Steeds (1969). For an overview of the history up to around 1980, see Sciberras and Payne (1985).

4 Mr. Passemard, Renault Automation, Industry Expert Interview Series (2000-2001).

5 Another example is de Figueiredo and Kyle (2000).

nies lost their market positions, went bankrupt, were taken over or left the market. When technological change occurred, the US producers' market for machine tools was in turbulence - from 1971 to 1986, only 5 of the 15 largest companies were able to maintain their position in the top 15[6] (see figure 4.1 below).

Figure 4.1: US machine tool industry - turbulence in the market

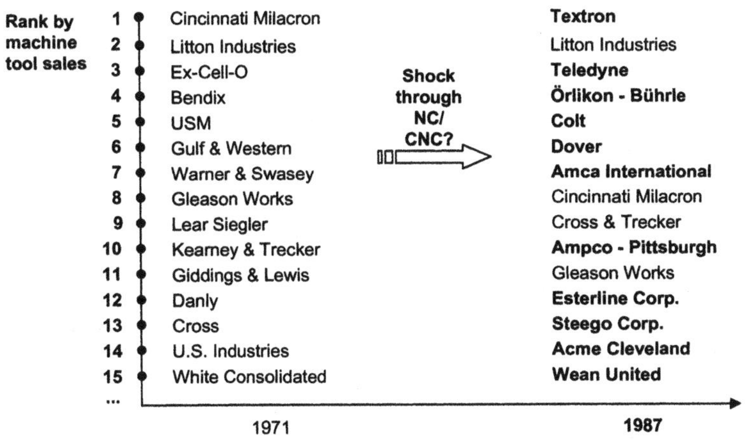

Legend: Normal type - established companies in the top 15, 1971
Bold type - new companies in the top 15, 1987
Source: author's analysis of data from Special News Bulletin, by A. Ashburn, American Machinist 1972 - 1988.

Between 1972 and 1987, the US lost its leading position as the largest producer of machine tools to Germany, the new entrant Japan (see figure below), and the Soviet Union as well (Fawcett et al., 1976). Japan gained the leading position in machine tool production in 1981.

At the same time, the British share of the world market fell from 8% in 1971 to 5% in 1977, and to 3% in 1981 (Production, 1983); employment fell from 81,000 in 1960 to 62,000 in 1976. Machine tool imports were grew 30% to 61% of domestic consumption between 1971 and 1982 (Arnott, 1983).

On the other hand, Swiss companies were able to retain their share of the world market despite the crisis they experienced in the watch industry[7], their most important home market. European companies generally declined in importance, however.

6 Ranking by revenue from Ashburn (1972-2000).

7 Lukas Weiss, Tornos Bechler, Industry Expert Interview Series (2000-2001).

Figure 4.2: Development of world market shares at the time of technological change

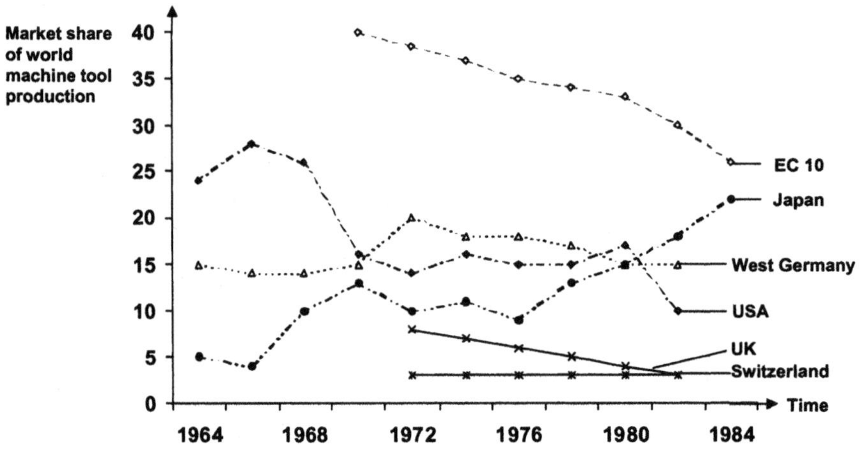

Source: own depiction based on data from American Machinist, Feb. 3, 1984, p. 2, Metalworking Production, and own estimates.

After 1973, the economies of the largest industrialized countries were characterized by slow growth. Users of machine tools were looking for ways to increase the efficiency of their operations. NC machine tools, which offered improved flexibility and reduced costs, were very welcome in this environment.

Data on worldwide competition at the single company level has been available since the late 1980s (Ashburn, 1972-2000). It indicates that the subsequent phase of technological change, namely the implementation of PC technology, led to similar types of changes in the market. Twelve years later, none of the companies that originally led the industry in 1971 remained in the top 15 (see figure 4.3 below). Two of the new leaders in 1999/ 2000, UNOVA and Thyssen Krupp, emerged from takeovers of several of the top-ranking companies in 1971. As a corporate investor, UNOVA owns two of the former top companies, Litton Industries, and Cincinnati Milacron. The German company Thyssen Krupp took over Cross&Trecker, Giddings&Lewis, and several other international companies, most within Germany[8].

[8] e.g. Witzig&Frank, Hüller Hille, etc.

Figure 4.3: Global machine tool industry - company ranking after CNC and during PC introduction

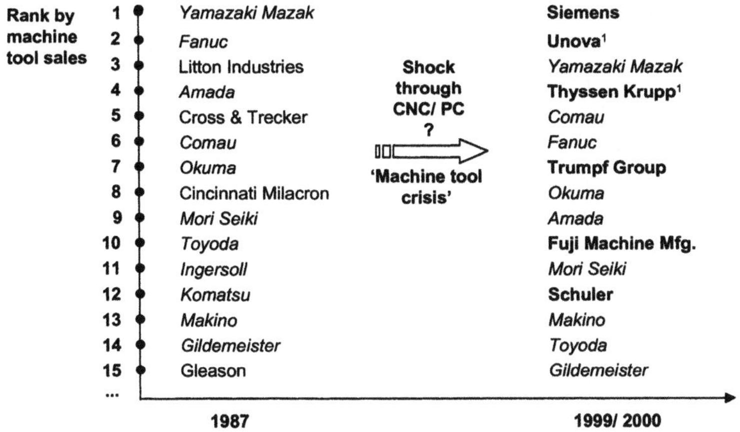

Legend: Superscript 1 - Unova emerged from Litton and Cincinnati, Thyssen bought Cross
Normal type - established companies already in the US top 15, 1972
In Italics - non-US companies in the global top 15, 1987
Bold type 1999 - new companies in the global top 15, 1999
Source: author's analysis of data from "The Blue Bulletin" by Anderson Ashburn, American Machine Tool Association, 1988 - 2000.

At first glance, the events within the industry might also seem to be the result of consolidation through mergers and takeovers. On the contrary, the total number of companies worldwide actually increased over the last 50 years[9]. The small-scale structure of the industry, which might be an indicator of market maturity[10], has basically remained unchanged from the 1970s until today (Marx, 1979) - the distribution of company size has remained the same since 1955. In Germany, 40% of these companies had less than 50 employees. At no time did more than 5% of the companies have more than 1,000 employees[11]. In fact, scale effects on the production side[12] seem to be rather weak. The variety of machines offered today is surprisingly large; almost no two machines seem to be identical. Each producer adds its

---

9 With established companies in the US and Europe disappearing and new companies entering in Japan.

10 Level of concentration expresses the degree of maturity in markets. See, for example, Klepper (1996).

11 Data taken from p. 9 of Schwab (1996) and p. 38 of VDW (1999b).

12 On the finance side, size did indeed have a positive effect on company survival. Especially in a hugely cyclical business with order volumes that can sometimes amount to 10% of the annual turnover, a strong credit line is necessary. In other words, implementing the new NC technology meant a huge financial investment. Mr. Willi Gauch, Industry Expert Interview Series (2000-2001).

own 'flair' and defines its own market niches (DeJong, 1998). The fact that many small companies operate in the machine tool market led to a phenomenon which is typical for this industry - the disproportion in the mere size of machine tool companies and customers. Consolidation in the automotive industry (as well as in the electronics industry) in the last 20 years puts machine tool producers in a position of low bargaining power. Especially when supplier and customer are one and the same big industrial giant (e.g. Siemens), small companies are in a particular weak position[13].

These special aspects of the machine tool industry and the events in the market accompanying and following technological change suggest that a closer investigation into the effects of the introduction of numerical control (NC) would be fruitful.

4.3 The availability of information about the industries' recent history

A new generation of decision-makers has recently moved into position in the machine tool industry. Most of the managers who witnessed the fundamental change of technology are retired or are about to retire shortly. Unfortunately, only a few studies that describe the history of innovation management within this industry have been conducted systematically (Mazzoleni, 1997). Researchers interested in business prefer to study industries with shorter reinvestment cycles. Results in such industries are evident in a shorter period and time series observations do not have to include intervals longer than a decade. The computer industry and related industries, which have cycles of less than five years, are often termed the 'fruit-flies' (Christensen, 1997) of business research.

This is not the case with the machine tool industry. The impact of NCs increased from the 1950s to this day. A present-day survey of the machine tool industry had to be conducted to be able to use primary resources and prevent the experiences of the former generation of managers from being lost. The research in this paper is based on qualitative interviews conducted in person or over the phone with over one hundred industry experts and decision-makers with an emphasis on motivating them to participate in the collection of data based on structured questionnaires[14]. Companies were selected from nine countries (Japan, Germany, Italy, Switzerland,

[13] e.g. when the terms of payment are not complied with. Mr. Cuatto, Giana, and Mr. Passemard, Industry Expert Interview Series (2000-2001).

[14] The list of participants in the interview series is found in the appendix to this chapter. I would like to express my sincere thanks to them.

Taiwan, United States, United Kingdom, France and Spain), covering the top nine machine tool exporting countries in 1999 (VDW, 1999a). Decision-makers from two[15] of the four large machine tool 'combines' in the former German Democratic Republic (GDR) were also interviewed. The majority of companies was established in the 1960s before the introduction of NCs upset the industry. It was necessary to use global sources in the scope of this study. What might seem valid on a national scale is not necessarily true when events in other countries are taken into consideration[16].

Comments that were made in the questionnaires were also included. In addition, statements from industry experts in academia, industry associations, and management consulting companies are contained in this survey.

This chapter builds on the extensive amount of relevant literature available It aims to fill in the gaps and perspectives that have not yet been covered in publications[17] while cross-checking the sources. The primary objective is to describe the process of digital control innovation in the machine tool industry and to confirm the disruptive nature of this new technology. It summarizes the history of the invention of NCs, discusses the diffusion of the technology in three major waves and elaborates on the extent to which NCs have influenced competition within the industry. A separate section explores the potential effects of other external factors during the 50-year period as well as Japan's remarkable success story. The numerical control technology and the categories and components of machine tools that shaped the industry structure are described in the appendix to chapter 4.

---

[15] Interviewed were managers from the combines '7. Oktober' and 'Fritz Heckert'. The other two combines were 'Schmalkalden' and 'Herbert Warnke'.

[16] For example the effects of the oil crises of the 70s left their mark on the diffusion of technology in the US industry, but not at all on the Japanese industry. The same is true for the number of firms - what might appear to be a major decrease in the number of companies from the German perspective does not hold true on a global scale, as quite a few companies in Spain, Taiwan, Japan entered the market at different periods of time. A typical essay from a German perspective is Schwab (1996); for the effects of concentration in Germany, see Fleischer (1997).

[17] Especially with regard to innovation management.

4.4 Location of invention, role of national research programs, and no advantage for first movers

4.4.1 The first years of a new technology

The modern development of NCs started around 1948 - 1949. A relevant stakeholder in this development was the United States Air Force, who wanted to have machines for the aircraft industries that were able to manufacture complex parts cheaper, faster and more accurately than by conventional methods. The Air Force then financed proof-of-feasibility studies of machine tools which could perform automated contour cutting. Under the leadership of John Parsons Company and the Servomechanisms Laboratory of MIT, the first experimental models of milling machines were developed and were tested in a few plants by 1951.

John T. Parsons, Vice President of the Parsons Corporation, Aircraft Division, in Traverse City, Michigan, conceived the initial idea of 'data systems' for machines. In spring 1949, he contacted Professor Gordon S. Brown of MIT's Servomechanisms Laboratory[18]. The Servomechanisms Laboratory[19] had conducted research on digital feedback control for military applications (control of gun turrets, etc.) during the war and was looking for industrial applications for their know-how. Research for potential application of digital feedback technology in machine tools had already been conducted[20]. Parsons was able to convince the Air Force to finance a proof-of-feasibility study. He envisioned the development of the 'Cardamatic Milling Machines I and II' in a joint effort by MIT and Parsons Corporation. The Parsons Corporation then received 200,000 USD[21] from

---

[18] A detailed account of the research project at MIT and the collaboration between MIT, the US Air Force, and the Parsons Corporation from MIT's perspective is given by one of the project managers in his book. See Reintjes (1991). Much of the information summarized here is taken from this book. Another summary of the first years, which differs with respect to certain points, is provided in Mazzoleni (1997).

[19] The Servomechanisms Laboratory was established in the late 30s and changed its name in 1959 to Electronic Systems Laboratory to account for a broadened scope of research.

[20] A PhD thesis, 'Analysis and Design of Sampled Data Systems' by William K. Linvill, was the result of a computer project at the MIT Department of Electrical Engineering called 'Whirlwind'.

[21] This amount turned out to be insufficient and was more than doubled in the course of the project. At this time, the annual salary of an engineer was about 5,000 USD.

the Air Force divided into four installments, of which 80,000 USD were used for research at the Servomechanism Laboratory. MIT alone spent a total of 360,000 USD on the development. The remaining 280,000 USD were billed directly to the Air Force. A conventional commercial three-axis milling machine from the Cincinnati Milling Machine Company was used as the basis for the experimental machine. The resulting prototypes were demonstrated publicly in September 1952.

After having sponsored the development of an experimental NC machine tool, the Air Force hoped that the industry would incorporate the new technology. Colonel Wilbur R. Carter, who then headed the Manufacturing Methods branch of the Air Materiel Command, urged the machine tool builders to turn the new technology into marketable products. The industry's reaction was far from enthusiastic. MIT and the Air Force even initiated a program for 'information-dissemination' of NC technology between 1953 and 1954.

4.4.2 Machine tool manufacturers get involved - what happened to the pioneers

Giddings&Lewis Machine Tool Company in Fond du Lac, Wisconsin (G&L) was one of the few exceptions that tried to improve the tolerance of its milling machines through numerical controls. Concerns about the complexity of the director unit (a simpler processor unit) led to another research project conducted together with MIT[22] and General Electric. With support from staff members from the Servo Lab, Giddings&Lewis spun off a subsidiary to produce digital directors, namely Concord Control Inc., Brighton, Massachusetts. The first commercial model of a five-axis NC milling machine went on display[23] at the Giddings&Lewis facilities[24] parallel to the National Machine Tool Builders Association Show in 1955.

Cincinnati Milling chose to establish a relationship with the British company EMI (Electrical and Musical Instruments) for the manufacture of control systems. This cooperation failed and Cincinnati began to produce control units of its own.

---

[22] The program's research supervisor was J. Francis Reintjes, who documented the research activities at MIT in his book. See Reintjes (1991).

[23] Much of the information in this paragraph was taken from Romeo (1975).

[24] The development team did not finish the machines in time, so G&L flew in interested parties from the exhibition.

In 1956, Kearney&Trecker manufactured one of the first commercial NC milling machines in the US[25]. The most famous name of one of these early models was 'Milwaukeematic'. The company cooperated with Bendix in Detroit, Michigan, as its supplier of electronic equipment and supplied products to the Aircraft Manufacturer Northrop Company. Bendix' Research Division had begun work on numerical control systems in the early 1950s, a project which was led by one of the Servo Lab's alumni. Kearney&Trecker as well as Giddings&Lewis[26] were later taken over by the German group Thyssen Production Systems[27].

One of Kearney&Trecker's first customers was Hughes Aircraft Company, Los Angeles, California. At that time, Hughes Aircraft Company employed about 30,000 workers and had divisions for electronics, aerospace, and defense. The visionary Howard Hughes[28] led the company. Around 1957, Hughes Company decided to develop a new product, a numerical control unit, and market it together with a machine tool (Raiser, 2001). Hughes had bought the first NC machine from Kearney&Trecker but decided to exploit the favorable exchange rate of the US Dollar abroad. With an exchange rate of 1 USD : 4.20 DEM and a relatively low interest in digital control technology on the part of the US machine tool industry, it is not surprising but little known that the mechanical side of one of the first[29] commercial machining centers was developed and produced by a German manufacturer: 'Burkhardt&Weber' in Reutlingen, Germany. Hughes Tool Company provided control units - which then consisted of radio tubes - as well as all other electric and hydraulic parts. Burkhardt&Weber designed and manufactured the machines. According to the agreement between the partners, Burkhardt&Weber was granted sales re-

25 Ernst Raiser, Ex-Burkhardt&Weber, Dzems Bruvelis, UBM, Industry Expert Interview Series (2000-2001). This incident is still used proudly in Kearney&Trecker's marketing material.

26 Giddings&Lewis went public in 1982, was bought by the American company EMCO in the late 1980s, and was finally taken over by the Thyssen Production System group in 1997.

27 Hüller became a key element of Thyssen Production System's chip-making machinery business in 1975. Over time, Burr special machines, Hille machining centers, Diedesheim special machines, Hessap lathes, Witzig&Frank special machines, Cross special machines, and Giddings&Lewis machining centers were added to the product portfolio. The respective companies were then taken over. Hans Ulrich Jaissle, Ex-Hüller, Industry Expert Interview Series (2000-2001).

28 Born to an upper class family in Texas, he owned 20th Century Fox as well as Trans World Airlines.

29 Some of the interview partners claimed that it was the first.

responsibility for Western Europe with the exception of Sweden, Hughes was responsible for the rest of the world. The NC machine series were named MT3 and MT3 A. These machines were based on classical drilling machines onto which the NC module was mounted. At the Hanover Exhibition in 1960, the machining center was presented to the public and attracted great attention but also great confusion, as the area of application for such machines was not clear to the visitors. Although Burkhardt&Weber took the technology seriously - they established an organizational unit for NC machines, which was about 30% of the size of the total company - Burkhardt&Weber did not benefit from first mover advantages. It was not until 1965 that the first MT3 A could be sold (to the French company Snecma, Paris). In the US, about 150 machines were sold by the mid-1970s. In 1966, Burkhardt&Weber started its own development program for a machining center, the MC 4, and developed a new concept for it. In the case of Burkhardt&Weber, NC technology did not lead to any considerable sales growth or share in the market. The company never grew out of its sub-critical size to finance its larger projects (some of them comprised up to 25% of the company's annual turnover of about 200 million DEM). In 1982, the company was bought by the Swiss technology corporation Georg Fischer, which sold it again in 1994. In 1992 and 2001, Burkhardt&Weber filed for bankruptcy and was recently taken over by the shareholders of the Italian corporation Riello[30].

As the value of the new technology and the potential competitive advantage it entailed were not fully appreciated in the US, the Japanese company Yamazaki was able to acquire the US license for the MT3 A from Hughes Aircraft. Yamazaki initially wanted to cooperate with Burkhardt&Weber. After this attempt failed, but after the failure of such a cooperation it they used the license to study NC machines[31]. Yamazaki had only built conventional lathes originally. In February 1966, the technical team at Yamazaki started to investigate new technologies, in particular NCs. They used the license to build one MT3 A machine and study the technology of machining centers. In the fall of 1968, Yamazaki was able to present its own series of NC machines - the turning centers 800R, 1000M, and 1500R, and the machining center BTC5 (Kuba, 1989). The access to NC technology was the basis for the company's later success as the market leader in machining centers, Yamazaki Mazak.

---

30 Dr. Hans Klein, Ernst Raiser, and Wilhelm Gauch, Ex-Burkhardt&Weber, Industry Expert Interview Series (2000-2001).

31 Dr. Hans Klein, Ex-Burkhardt&Weber, Industry Expert Interview Series (2000-2001).

In Germany, Schiess AG[32] presented the first heavy milling and boring machine in 1957 using a Brown, Bowerie and Cie. controller (Fleischer, 1997). The company merged with Dörries Scharmann in 1992 and went bankrupt in 1994, the same year its parent company, Bremer Vulkan filed for bankruptcy. In 1959, Pittler AG produced the first German NC lathe (Glunk, 1989; Schwab, 1996)[33].

The period between 1959 and 1965 was a time of rapid expansion for numerical controls. The first multifunction, multipurpose machining center was introduced in 1958 (Rendeiro, 1985). At the 1960 IMTE show in Chicago, already 90 different models of NC machines were presented.

4.4.3 National programs for developing a new technology

The need for a new generation of sensors that could express measured velocities and positions in digital form emerged with the development of automated machines. At the same time, a market was created for a digital computer that could be integrated into sensors, controllers, and machine tools and manage the real-time digital control of the entire system. There was an opportunity for a breakthrough in the concept of programming. The idea was to create hierarchically-structured computer languages in which commands could be expressed in simple English-sounding words rather than cryptic, codified form that would also be able to automatically compile subprograms into an integrated systems program. 'The need propelled the research', as one of the MIT program managers, Dr. Reintjes, put it (Reintjes, 1991).

Most industrialized countries started national programs to meet these needs and to accelerate and intensify the development and introduction of NCs.

In Germany, five universities - Berlin (Prof. Spur), Stuttgart (Prof. Ehrhardt, Prof. Stute), Aachen (Prof. Opitz), Karlsruhe (Prof. Warnecke)

---

[32] Dr. Reinhard Fleck, Ex-Schiess AG, Industry Expert Interview Series (2000-2001).

[33] 14 German companies showed NC machine tools at the 1960 Hanover exhibition. These included Berliner Maschinenbau AG, Burkhardt&Weber, Bohle, Collet&Engelhard, Droop&Rein, F. Werner, Heller, H. Kolb, Hüller, Pittler, Scharmann, and Waldrich. The overview on pp. 530/531 in Spur (1991) is inconsistent with other sources like Schwab and Glunk, as well as p. 200 in Fleischer (1997) with respect to which machine types these companies presented. The interview series with industry experts named Collet&Engelhard, Pittler among others as producers of lathes around that time. The overview in Spur categorizes them as producers of boring mills.

and Hanover (Prof. Kienzle, Prof. Osenberg) cooperated closely with industry and industrial organizations. These universities provided specialized knowledge on individual manufacturing functions such as turning, milling, forming, computer control, and computer design.[34] As a result, several direct numerical control systems were developed, most notably a national manufacturing software system called 'EXAPT', which is still being promoted by the 'EXAPT-Verein' in Aachen, Germany. These institutions also trained management resources for the industry. Many industry experts who participated in this study as well as contributors to the survey can be found in the Aachen alumni register (Eversheim et al., 1980). Many industrial organizations, notably the 'Verein Deutscher Werkzeugmaschinenfabriken e.V.' (VDW) helped identify market needs and tried to initiate collaborations and research projects. The effort to start horizontal cooperations was largely unsuccessful as most of the companies preferred to conduct their business alone, but several consulting reports delivered insight into the state of the art and competitive environment. It was, however, different for vertical cooperations, traditionally of importance in Germany[35]. Companies cooperated strongly either with their customers or, more often, with their supplier of numerical control units, namely Siemens. Siemens began developing numerical control units in 1958 and was able to deliver the first units two years later (Schwab, 1996). The electronics company helped to bring the German machine tool industry up to speed and organized training programs for machine tool companies.

Similar steps were taken in Norway, where four of the country's leading firms cooperated with the Technical University of Norway's Central Institute of Industrial Research in order to develop a national manufacturing software system.

France's association of the fifty largest manufacturers assisted in the acquisition of high-tech equipment, management and technical information and education for managers.

In the Netherlands, university research focused on the development of software for automatic production processes.

In the United Kingdom, the efforts of the Production Engineering Research Association, universities, and industry groups focused on cellular manufacturing in order to reduce production time. The industry, though, was technologically somewhat 'backward' (Daly et al., 1980) and too slow in picking up CNC technology as a comparison with the Swiss industry

[34] Taiwan has imitated the approach of university – industry cooperation for R&D in recent years; see Hsu and Chang (2000).

[35] See the discussion of results from the 'Mannheim Innovation Panel' in Harabi (1998).

showed (Ackermann et al., 1985). The Swiss industry invested substantially in R&D to innovate their product performance. Through the timely adoption of new technologies, they were able to remain in a leading position in the world markets, whereas the UK industry, with a few exceptions[36] (mostly the '600 group'), almost disappeared.

In particular, Japan, which will be described in a separate section below, and Germany developed the ability to create completely automated systems.

As US companies fell further behind and after more than 200 companies had dropped out of the market by 1984, the necessity to spur research and development efforts was felt. Measured as a percentage of the GNP (gross national product), Japan and Germany both spent more in private research and development than the US (Anonymous, 1980). Leading manufacturers like Cincinnati Milacron increased their R&D budgets; several industry groups formed joint research ventures while the automotive industry, as the most important client industry, the Department of Defense, and individual states provided financial backing.[37]

4.5 Patterns and extent of technological change: between shock and evolution

Over the course of fifty years, three major technological innovations had a major effect on the machine tool industry. Three waves, each lasting about ten years, can be observed during which the new technologies experienced strong diffusion and were implemented within the manufacturing industry.

As discussed[38] in chapter 1, the extent to which the new technologies appeared as a major disruption or a shock to the industry can be expressed in three categories. The first category is a fundamental change in product architecture. Numerical controls not only resulted in the innovation of one component, it meant the complete redesign of the whole machine and its implementation into an automated production system. Several machine tool groups, each representing one specific function, were converged into multi-purpose machines.

---

36 Dr. Stephen LeBeau, 600 Group, Industry Expert Interview Series (2000-2001).

37 Cincinnati Milacron increased its R&D budget from 1983 to 1984 by 19% to 36 million USD. General Motors decided to increase research on advanced manufacturing research, and the government spent about 37 million USD. See Unger (1985). A US perspective is taken in Forrant (1997).

38 Archibugi and Simonetti (1998) present an alternative explanation.

The second category was a disruption in competence, which gives companies competitive advantage. The competence to manufacture highly accurate mechanics made way for the need to develop electrical engineering know-how. Numerical controls enabled most manufacturers to produce accurate machines. A change in the value chain occurred; numerical controls were mostly purchased from electronics companies.

The third category is that of value proposition. Numerical controls and sensors reduced the value of the image that established companies had built around themselves, i.e. that they would be exclusive producers of machines with high accuracy. Now almost every manufacturer could increase the accuracy of his machines with the help of sensors and NC. The criteria for the purchase of a machine tool became different. Price was among one of the most important factors. It indicates a change in how products were valued.

4.5.1 Three waves of technological innovation

To many contemporary decision-makers, the change in technology and its effects on the companies' success appeared[39] to be gradual. Long reinvestment cycles of 10-20 years[40] in the manufacturing industry were taken as given because machine tools last a long time[41]. With rather long reinvestment cycles, even drastic technological disruptions were leveled out at first sight over the years[42]. But when longer periods are taken into consideration, a wave pattern expressing how the different NC technologies affected the industry can be observed. From today's perspective, the development and implementation of the new technology happened in three waves.

39 Dr. Klein doubts the disruptive effect of this technological change. He observed a fundamental, although rather incremental, change in the industry. Industry Expert Interview Series (2000-2001).

40 'Long waves' of about 10 - 20 years at the company level have been observed in Klump and Männel (1995) The conclusion here is also drawn for the manufacturing industry in general.

41 With appropriate service, maintenance, and occasional reconditioning they can perform for many years. Between 1925 and 1950 about 50% of the machine tools used in the US were 10 or more years old. Roughly 15% were over 20 years old. See p. 61 in Wagoner (1966).

42 This is the reason why researchers often use data from the computer industry and other related industries. Short reinvestment cycles make strong influences easily visible after a short period of time. Some researchers have therefore termed this industry the 'fruit fly' of business research.

The **first wave** was obviously the introduction of numerical controls for machine tools in the 1950s and 1960s. From 1959 to 1965, the number of shipments increased over a seven year period by 40% on average[43], i.e. from about 200 to about 2,050, which then stagnated over the following years. Customers were primarily from high-end manufacturing industries like the aerospace industry, which had driven development from the beginning. The total value of shipments kept growing at this pace even after 1965[44].

The **second wave** in the 1970s and 1980s was triggered by the demand for microcomputers in numerical controls. Sales of NC machines grew dramatically by about 30 - 40% annually from 1978 to 1984.

Like the other two waves, the second wave was triggered by major price cuts[45] and performance increases in the processor market. The price for a minicomputer CNC unit in 1973 had dropped to roughly 50% of that of an NC unit in 1963. The price of a microprocessor CNC unit in 1978 was only about a quarter of the price of a NC unit, and in 1982, the price for a multiprocessor CNC unit was even lower.

Available US data makes it possible to plot the increase in machine stock in one of the largest machine tool markets (see diagram below) for the period of the first and second wave. The curve shape resembles an 's' with inflection peaks in 1967, 1979 and in the mid-1980s. The first wave ebbed and initial fascination about the new technology decreased around 1969. By that time, the market niches for the application of hardwired NC machines had been occupied.

After 1967, there was no significant increase in demand until 12 years later when CNC machines were available in large numbers. These products offered new features and were far more flexible. By separating software from hardware, the same control circuits could be built into different machines. In the predecessor versions, every machine required its own type of control circuit. Now controls could be customized for different machine types by simply loading a new software program. In the same way, adapting to new shapes did not require exchanging the set of punched tapes and feeding new ones. The machines were prepared for new shapes through activating a different subroutine. The increase in flexibility and the drop in price made CNC machine tools interesting for the majority of manufacturing industries. This interest expresses itself in another strong wave of demand, which lasted for over 5 years and ebbed off after 1984.

---

[43] Compound Annual Growth Rate (CAGR).
[44] Data is taken from USDoC (1959 - 1968).
[45] See graph p. 553 in Spur (1991).

Seen in absolute numbers the second wave was much stronger than the first wave due to three main developments that occurred simultaneously. The first development, as discussed above, was the emergence of a mass market. 'Mainstream' producers became interested and digitally controlled machines left their niches. Starting in 1980, digital control was also introduced for special machines[46] as well as for forming machines[47], which had mostly been mechanically controlled until then. Few areas of chip-making metalworking, like gear cutting and threading, were left in which digital controls had not yet dominated. For these machines, which did not require a lot of flexibility to begin with, the integration of NC technology took place after a period of several years. This delay is partially responsible for the length of the waves. In other words, when looking at the diffusion rate for one single machine type, the wave pattern for each category would have been even more obvious despite the shift in time.

The second development was around 1980 when Japanese companies started a massive invasion of the export markets, especially the US market. Section G of this article elaborates on Japan's success. By focusing on the highest economic gains for their customers, Japanese companies were able to capture a large portion of the world market. An ample supply of CNC machines that undercut the price of their Western competitors by over 50%[48] contributed extensively to standardization and thus to the availability of moderately-priced CNC machines.

Finally, the third development was triggered by the effect computers had on the entire value chain[49]. They enabled the transition from paper-based to electronic work in the drawing office (computer-aided design - CAD) and led to efforts to support production with computers (computer-aided manufacturing - CAM) (Feigenbaum, 1980). The ultimate goal was to integrate and automate the entire production process (computer-integrated manufacturing - CIM). Some industry experts call it the 'input of raw material at the beginning of the production line and output of the finished product at its end'.

---

46 Mr. Hölzl, Ex-Hüller, Industry Expert Interview Series (2000-2001).

47 Mr. Passemard, Renault Automation, and Herr Brandstetter, Schuler Pressen, Industry Expert Interview Series (2000-2001). Especially the extended machine environment, like the integration of automated work piece/ pallet-changing systems and tool changers, requires the use of numerical controls.

48 See price ratios on p.123 of Ehrnberg and Jacobsson (1997).

49 Afuah expanded Henderson's concept for architectural innovation to the whole value chain. He expects that technological change that influences the conceptual design of products will have its influence throughout the value chain. See Afuah and Bahram (1995).

Figure 4.4: Diffusion of NC machines - growth in stock

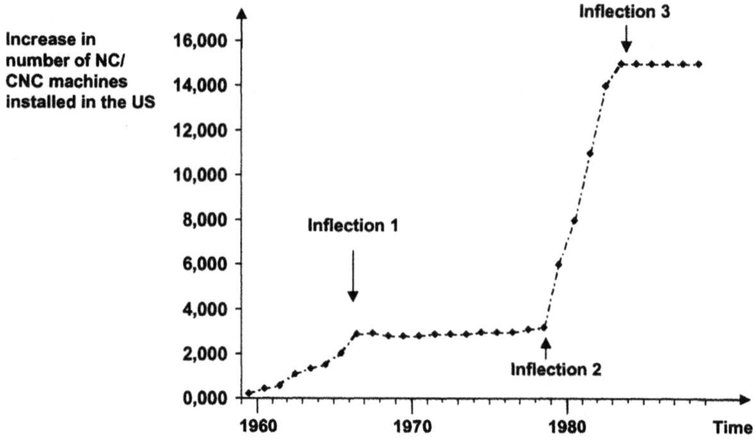

Source: own depiction, based on data from US Department of Commerce, 1959 - 1968, after 1968 estimates based on the American Machinist inventory of stock increase 1973, 1978, 1983 taken from Ehrnberg, Ellinor, and Staffan Jacobsson, R & D Management 27, no. 2 (1997): p. 123.

'Investigations in Computer-Aided Design for Numerically Controlled Manufacturing Processes' was the title of a study at MIT, financed by the Air Force from late 1959 to 1970 with over 3 million USD[50].

Many of the national programs mentioned above, like the EXAPT society, aimed at promoting CIM, but the problems soon appeared to be too complex to be solved entirely by computers (Merchant, 2000). This first integration with actual production machines was not successful on a broad scale. Data handling was too complex and the systems too rigid. Using reduced volumes of data, the efforts led to the implementation of production planning software, called PPS for production planning systems. CNC units were built from hardware and software that were proprietary systems and comparatively difficult to handle; special training was still necessary.

The personal computer (PC), which started its triumphal march after a period of experimentation (Ceruzzi, 1996)[51] in the early 1980s, meant controls that were standardized, cheaper, and easier to handle. The **third wave** of technological implementation continues into to the present. PC-based CNC appeared on the scene around 1990. The first models were too restricted in processing power and memory to handle the data volumes re-

[50] For a detailed description, see p. 94 - 133 in Reintjes (1991).

[51] For a historical account of the phase of technological fervent in the PC industry, see Ceruzzi (1996).

quired for sophisticated machines. They could provide low cost alternatives for CNC machine retrofits at the most. By 1994, these hurdles were surpassed by controls' manufacturers like Siemens and Fanuc. PCs were embedded into CNC units (Chalmers, 1998). The more hardware-oriented older programming languages were gradually substituted by graphical user interfaces (GUI).

The third wave is characterized by yet another change in the competitive landscape - German manufacturers regained ground in international competition while other European manufacturers such as Italian and Spanish companies are playing a more important role than ever before. In general, most producers, even small ones, must be present in the international market. Technological development splits the market into two segments (Kawakami et al., 1985) - the first segment, high-efficiency production; the second, ultra-modern technology (e.g. ultra-high speed, ultra-precision). Companies from Taiwan (Jones et al., 1984), Korea, and China[52] are trying to set foot in the high-efficiency segment, in which Japanese manufacturers were able to establish a dominant position.

A diagram that provides an overview of the three shock waves that were caused by digital controls is displayed below. It depicts the use of the various digital control technologies in vital segments of the production value chain - the design or drawing office, machine tool production, and the use of the machine tool in the production process. The pointed end of each wedge represents the starting point of each technology. The diameter of the wedge illustrates the level of diffusion in its major areas of application.

The three waves of diffusion within the market are marked in gray. The first and second waves are located between significant inflections in the rate of diffusion as illustrated above using the example of the US market. The interview series revealed that the European market followed quite a similar pattern with respect to time.

[52] By 1981, China's machine tool industry was already able to supply 99% of China's requirements of machine tools. Machine tool exports began at the end of the 1970s. In 1979 and 1980, China exported 10,000 machine tools. See Xunxian (1981).

Figure 4.5: Three waves of digital control technology

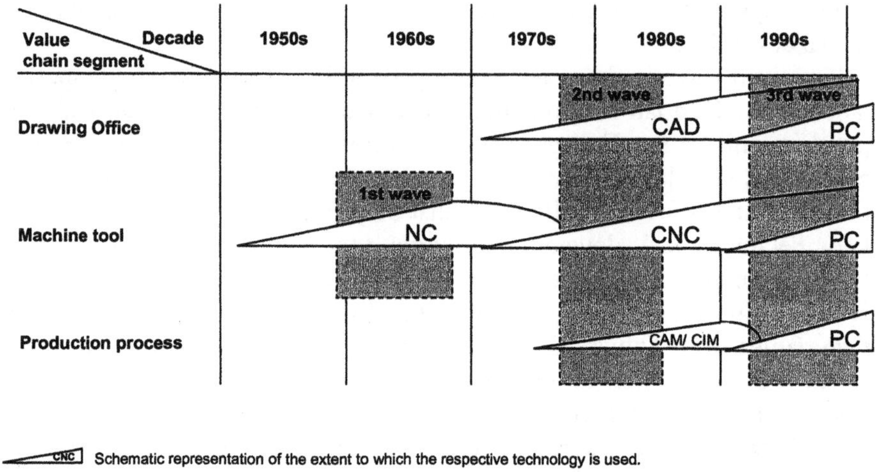

Schematic representation of the extent to which the respective technology is used.

Source: own depiction, based on discussions with Dzems M. Bruvelis, UBM, and Wilhelm Gauch, Alfing and various literature.

Among different manufacturing industries, however, the rate of diffusion varied considerably. The level of concentration in the respective customers' industries also seemed to influence diffusion. The number of different types of NC machines sold to manufacturing industries during the three phases in general increased respectively over 20 to 30 years. This becomes evident when looking at the adoption rate for selected industries. The time frames from the first use of NC machines to the time when 25% of all machine purchases included NC technology can range between 6 and 15 years (Romeo, 1975). In the chart above, the wedges symbolizing the adoption of NC technology by the user industries span decades. The diffusion pattern of digital technology, numbers of the development of machine stock in the US as well as the hazard count in the survey of the 59 international companies are all indicators for a technology shock in three waves which is detailed in chapters 5 and 6. The interviews revealed that several smaller industry segments still undergo shocks caused by digital control technology ahead of them. Disruptions will potentially upset the competitive environment in these segments in a similar fashion as the three big waves did throughout the entire machine tool industry. For the sake of brevity, they are not elaborated on here.

4.5.2 Innovation of product architecture

In most cases, the numerical control unit was retrofitted onto existing machine concepts until the capabilities of numerically controlled machines were fully understood and the machines completely redesigned. The electronic control posed fundamental problems for mechanical parts. The digital control was able to maneuver the tools at a much higher speed than the mechanics allowed. It was quickly understood that innovation not only involved one component, the control unit, but the whole machine[53].

Numerical controls were able to improve or balance out the accuracy of mechanical parts. In addition, NCs enabled companies to manufacture products that they had not been able to manufacture previously. First of all, electronics replaced many mechanical functions (substitution). Then several functions were combined. It started with the combination of turning and other functions[54]. Numerical controls enabled the integration of several functions into one machine and started a process of convergence from function-specific machines to flexible machining centers (see figure below). In the 1960s, the functions of milling, drilling, planing, and boring could be combined into machining centers. The demand for versatile machines grew as lines in the production industry, e.g. the automotive industry, went from a small portfolio with few types in mass production to an increased type portfolio with smaller quantities. Later, as the precision of NC systems increased, the functions of grinding and turning could be combined. These developments allowed a convergence of turning centers and machining centers to appear realistic[55].

The combination of functions in one machine was carried out by changing the entire product concept. Numerical controls made it possible, not by putting together existing modules or improving existing machine types, but by understanding the NC capabilities and planning the machine architecture right from the start, i.e. the drawing board. According to Henderson's (Henderson et al., 1990) and Afuah's (Afuah et al., 1995) categorization, this indicates technological change of a large extent called 'architectural innovation'.

---

53 See 'The lessons learned', p. 167 in Reintjes (1991).

54 Mr. Cuatto, Giana, Industry Expert Interview Series (2000-2001).

55 Klaus Rietschel, Ex-Niles Chemnitz, Industry Expert Interview Series (2000-2001).

Figure 4.6: Convergence of cutting functionalities under the influence of NC

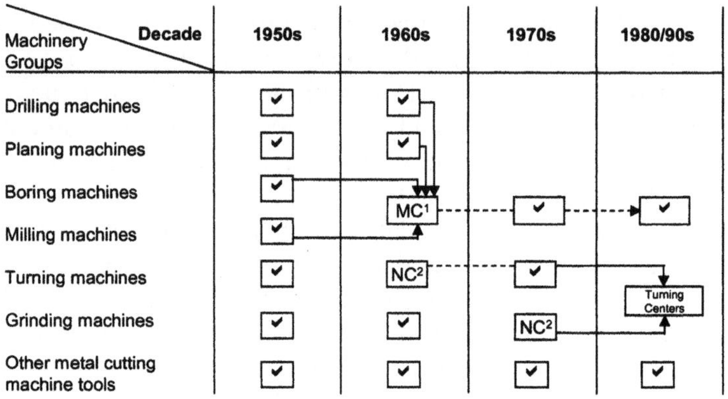

1) Machining centers, short MC. 2) Covers both NC and CNC.
Source: own depiction, based on discussions with Dzems M. Bruvelis, UBM.

With different functionalities converging, classical machinery group specializations for cutting machine tools disappeared to a large extent. In addition, manufacturers were expected to supply more than just the machine. Customers preferred to buy entire systems from suppliers, including peripheral products and services - fixtures, tools, set-up, training, maintenance, etc. Not only did convergence eliminate the borders between industry groups, it also led to a redefinition of market segments. As can be observed at recent industry exhibitions[56], producers today are divided into two large groups - the first group dealing with form-defining functionalities (drilling, planing, boring, milling) and the second group focusing on surface treatment (turning, grinding).

[56] At the EMO 2001 in Hanover, milling machines, machining centers, flexible production systems and robotics were presented together in the exhibition halls as one group; in the other halls, grinding machines and machines for honing, lapping, polishing, etc. were shown together; see EMO (2001).

4.5.3 The technology's disruptive effect on companies' capabilities and the loss of competitive advantage

Naturally the new technology did not affect all producers of machine tools in the same way. The use of numerical controls gave machines first of all higher flexibility. Above all, when workpieces are complex and require a number of different tool positions and movements, the programmable NC machine proved to be much more efficient than resetting the controls manually (Dreher, 1997), see figure below. In the same way, NCs proved to be superior when a large number of different workpieces were to be produced using one machine[57]. The situation for high quantities of the same workpiece, however, was different. Here, production in 'high-throughput' transfer lines - where durable mechanics are key - was more efficient. The advantages of NCs here were marginal[58].

The result of the diffusion of the new technology was that NCs were introduced at a much earlier stage for applications that required high flexibility. For a period of time, the machine tool industry was divided into two groups: companies that understood themselves as producers of machines for flexible use and those that delivered high productivity lines.[59] In recent years, the main customers for transfer lines, namely the automotive industry, increased the variety of their models and sub-models. Consequently, their production lines required greater flexibility. Since then the producers of transfer lines have also been under pressure to allow the lines to produce different workpieces.

Before numerical controls appeared on the scene, the capabilities of producing machines with highly precise mechanical components provided important sales arguments and a competitive advantage. There were even

---

57 A detailed discussion of different production principles is found in FhG-ISI (1982).

58 Some authors (e.g. Ehrnberg and Jacobsson, 1997) claim that there has been a discontinuity in the transition from the CNC machine to the Flexible Manufacturing System. They fail to see that both exist concurrently and are just solutions that differ from each other by the degree of flexibility and efficiency they offer. In addition they found another disruption in the development from conventional to CNC machines. This on the other hand is a claim which is a bit superficial, because it leaves out NC. But overall this second finding it is more along the line of what is observed in this research.

59 G. Moillet, COO of Mikron AG does not feel that the producers of transfer lines were greatly affected by effects of the NC technology. The control unit for high productivity lines does not require a great deal of flexibility but is still dependent on highly accurate mechanics. Industry Expert Interview Series (2000-2001).

national differences in the accuracy of precision machines delivered by the various companies (see figure 4.8)[60]. In the 1950s, German companies produced machines that met most accuracy requirements in the manufacturing industries. German producers were well known for their high-quality machines, especially in the automotive industry. This reputation helped them to regain a strong position in international competition and even to become the number one exporter of machine tools in the 1970s.

Figure 4.7: Overview of production concepts

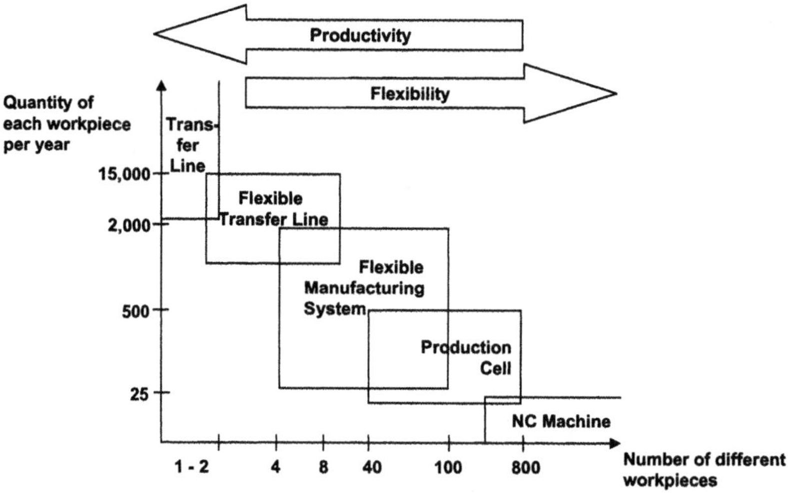

Source: own depiction, based on depiction p. 151 of Carsten Dreher, "Technologiepolitik und Technikfusion", Baden-Baden, Nomos Verlagsgesellschaft 1997.

Many Swiss companies, who were traditionally closely tied to the watch industry, focused on precision engineering. Their machines achieved accuracies suited for tasks requiring an extremely high quality surface and shape.

Manufacturers from the UK, USA, and Italy[61] had to make considerable efforts to improve the accuracy of their products.

Expert interviews were helpful in creating an index of estimates that describes how large the differences in accuracy were (see diagram below). If Germany would be the base with 100 points at the beginning of the 1950s, several Swiss companies responsible for the country's high-quality im-

[60] Dr. Klein, Ex-Burkhardt&Weber, and Prof. Weidemann, Ex-VDF and Ex-Pittler, Industry Expert Interview Series (2000-2001).

[61] Also Spain and France.

age[62], would be rated at over 120 points. Companies with some output problems, such as UK, US and Italian companies, would lie at around 80 points. Through increased efforts to enhance production procedures, conventional machines from these countries would have narrowed the gap between them and the leaders Germany and Switzerland.

The capabilities of probing systems and the adaptive behavior of CNC machines basically put most of the producers, including Japanese companies, on the same level of machine accuracy. Two techniques were instrumental for a quantum leap in workpiece accuracy. Firstly, geometrical shortcomings of machines could be corrected by entering correction values into the software. In addition, temperature caused workpieces to expand and machines could be corrected in the process with the aid of probing systems. The skill of producing highly precise mechanical parts and the knowledge of 'tricks' about how to deal mechanically with expansion due to temperature became less important. The new control technology had a major impact on the company competence[63] required in the successful machine tool market, dramatically devaluating the old competence accrued over generations.

Figure 4.8: Development of positioning accuracy influenced by CNC and in-process measuring

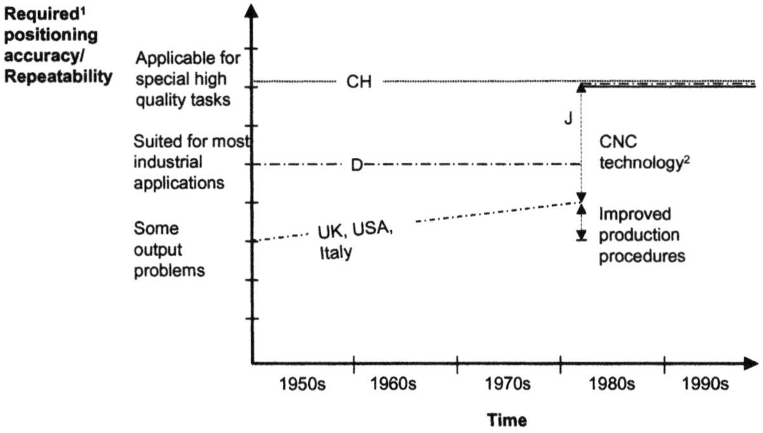

1 Qualitative, based on expert interviews.
2 With automatic feedback loop from measurement unit.

Source: own depiction, based on expert interviews and discussions with Dzems M. Bruvelis, UBM.

---

[62] Like the company Dixi Machines SA.

[63] For the importance of competence when technological change occurs, see Tushman and Anderson (1986).

Since machine tools were invented, it was the mechanical engineer who contributed critical knowledge to machine design. Experience in good mechanical design paid off, especially when it came to the durability and accuracy of products. The advent of digital control systems brought about a need for electrical and electronic competence within machine tool companies. The emphasis in companies not only shifted from specialized to standardized mechanics[64] but also from mechanics to electronics overall. Electrical engineering departments were created whose responsibility was either to modify standard numerical controls or, in addition, to develop specific controls and to plan the integration of machines into automated processes. In most cases, however, the result was vertical disintegration. Machine tool manufacturers no longer produced the control unit. Usually only so much competence was built up in-house as to permit the integration of digital controls into machines and the integration of the machines into automated production lines. At the 1965 Machine Tool Builders' Show in Chicago, it became evident that the value chain for machine tools had experienced a major change. Control units no longer fell into the machine tool builders' competence, they were mostly developed and supplied by electronics companies in the US such as General Electrics, Bunker-Ramo, and Superior Electric (Reintjes, 1991), later also by companies like Fanuc in Japan and Siemens[65] in Germany.

At the other end of the production chain, companies simultaneously decreased their depth of vertical integration due to cost pressure[66] and increased the percentage of mechanical components used. These parts were bought from cheaper, specialized suppliers, sometimes from low-earning countries.

[64] Mr. Richard Curless, Cincinnati Machine, and Mr. Rainer Nehls, Mercer Management Consulting, Industry Expert Interview Series (2000-2001).

[65] Detailed case studies on the development at MIT, at Fanuc, and at Siemens were performed by the working group 'Culture and Technical Innovation' at the Academy of Sciences and Technology in Berlin, and published in Spur, Specht and Schröder (1994) and Schröder (1995).

[66] Ulrich Vetter confirmed the fact that cost pressure led to vertical disintegration in mechanical parts: Industry Expert Interview Series (2000-2001). It was not that the increase in quality by suppliers had prompted machine tool companies to outsource. This was misunderstood in Rendeiro (1985). The increase in mechanical quality had taken place earlier (until the early 1980s), but within the machine tool companies themselves (see second figure in figure 4.8).

4.5.4 Change in the product's value proposition and order of magnitude improvement in value for users

For years, accuracy had been one important barrier to entry into the market. Buyers had extremely high brand awareness; established companies had the image of producing accurate machines of high quality. NC rendered this image practically worthless. Electronic controls now enabled most companies to manufacture machines that could produce at a level of accuracy that even exceeded was even above market demand. The decision criteria for purchasing a machine tool had changed.

Price inflation in microprocessors and electronic memory led to a drop in price for NC units (see section 4.5.1). This allowed Japanese companies that had focused from the beginning on the integration of computers in the control units of standard machines to offer CNC machines in high volumes at low prices. The Japanese companies could play out their price advantage[67] and successfully entered the world market. They even overtook Germany as the largest supplier of machine tools[68].

The amount of money customers were willing to spend on numerical machine tools is taken into consideration to quantify the advantages. Around 1970, customers on average spent five times more on a NC machine than on a classical mechanical one[69]. Towards the end of the 1970s the ratio was even more extreme. According to the US census, the average price for a NC machine was 190,000 USD, whereas the prices for a purely mechanical machine were between 5,000 and 50,000 USD[70]. The example of cylinder grinding machines illustrates this. The NC version would cost around half a million USD, whereas a purely mechanical machine for the same purpose would sell for only about 50,000 USD - a factor of 10. For a long time, the customer's motto was 'If you have a good product, you can afford an NC machine'[71]. Literature has also identified the price perform-

---

67 Dzems Bruvelis, UBM, Industry Expert Interview Series (2000-2001).

68 The Japanese approach was soon understood by the German industry. Consulting projects in the late 1970s initiated by the German machine tool association VDW and by large manufacturers like Deckel and Maho identified the scale advantages of CNC mass production of standard machines. One effort to create a German mass producer was the merger of Deckel and Maho which resulted in the failure of the new firm. Gildemeister, strongly supported by their financiers, took over the remnants in 1994. Sources: VDW, Dr. Heinz Drink, etc. in Industry Expert Interview Series (2000-2001).

69 Numbers from USDoC (1959 - 1968).

70 Numbers from USDoC (1959 - 1968) and Marx (1979).

71 Dr. Zoltan Szabo, Ex-Hüller-Hille, Industry Expert Interview Series (2000-2001).

ance ratio as one of the strongest indicators for technological change. Neither patents nor bibliometrics can give early warnings. These indicators rather follow than precede innovation and the subsequent diffusion process (Ehrnberg et al., 1997; Schmookler, 1962).

The increase in value also left its mark on the diffusion pattern of NC technology in the customers' industries. Large manufacturing companies replaced their machine arsenal more quickly with machines which had the new control technology, a trend which continues until today. The usage of NC machines in companies with more than 1,000 employees is more than double than in companies with 100 employees and less (Lee, 1996). The price of a NC machine might simply be too high for many small companies.

4.5.5 Technological change in the machine tool industry and the technology shock matrix

The events of the last decades in the machine tool industry described in sections 4.5.2 through 4.5.4 entail 3 major categories of technological change which can be summarized in the following way.

The first is the fundamental redesign of machine architecture: After the first trial versions, it became apparent that the capabilities of digital controls were limited by the conventional mechanical design. Originally, the digital control units had simply been mounted onto the existing mechanically controlled machines so far. Therefore, the new and successful generation of numerically controlled machine tools was built based on an entirely new concept. The new machines were not only fully equipped to meet the capabilities of digital control, over time they also integrated the machine environment such as raw material feed, in-process gauging, or automated tool changing (ATC).

The next category is the change in the perception of value: Excellence in mechanics - which meant accuracy and reliability of products - was the foundation on which the image of established companies was built. This image had posed a barrier to new entrants for decades. NC and the possibility of integrating in-process measuring offered a way to achieve high accuracy and extraordinary flexibility while at the same time replacing mechanical controls with electronics and software. It also meant order of magnitude gains in efficiency for the users. Customers were willing to pay a premium for the new flexible NC machines and also to accept products from new entrants who - like the Japanese - were competing through price. At first, established companies lost shares to the aggressors in the standard machines' sector - later also in the special machines market.

The final category is the dramatic devaluation of traditional competence: The shift from mechanical to numerical controls meant a dramatic shift in competence. Electrical engineers and programmers moved to the fore in companies which were formerly dominated by mechanical engineers. Once a core competence, the control unit was now - after some by and large unsuccessful efforts to produce them in-house - being manufactured by electronic companies such as Fujitsu-Fanuc or Siemens. The industry, grouped around mechanical specialization, was split up into three main groups of 'chip-making machines': machining centers, turning centers and grinding machines. Specialized boring and milling machines disappeared. With the control unit now being purchased from suppliers, a large portion of the added value migrated to the electronics companies. The control unit was then available to everyone - mechanical accuracy had lost its role as an entry barrier. A new battle for market shares had begun.

These categories, which had been derived theoretically in chapter 1, have been detected in the research object.

It was found that aligning the axes of product properties and customer perception is also justified for the machine tool industry: When customers' opinions about machines changed, priorities regarding essential performance parameters also shifted. While accuracy and reliability were important previously, this had now changed to price and flexibility. Accuracy and reliability were still considered a must, but purchasing decisions were more strongly influenced by price and flexibility. Improvements in the former performance categories 'accuracy' or 'reliability' would have been sustainable in character. An improvement in 'flexibility', however, was a novelty. It meant that customers perceived additional value in new categories. When improvement along a new dimension leads to a higher value of a machine for the customer, this technological change is disruptive in character. A change in perception is connected with a disruption in the conventional performance trajectory. Sustaining changes lead to continued growth in what is referred to as classic parameters. Disruptive changes break with the traditional trajectory by allowing higher ratings in other factors.

The performance of a machine does not usually change much unless through a fundamental redesign or an entirely new concept. Thus, the horizontal categories of 'performance trajectory' and 'innovation of product properties' usually express the same phenomenon and move in same direction.

The extent of threat to established businesses and the degree of opportunities for new ones depends on the product groups. For the press manufacturers, the availability of digital control expanded without substituting their core know-how. However, the critical competence still remained in the mastery of complex mechanical systems. Numerical control became significant in the course of supplying complete automated systems, including automated tool and workpiece exchange. For the chip-making industries, however, NC/ CNC meant gains in efficiency and flexibility. On average the price of workpieces dropped significantly; as it did similarly for milling, boring, turning, and grinding applications. As for gear cutting, development increased even further through NC/ CNC. Not only did prices per workpiece drop and flexibility increase, but digital controls also allowed for movements and processes which could not be achieved previously. Moreover, new shapes of workpieces became possible in gear cutting.

In this sense, the technology shock was felt most by manufacturers of gear cutting equipment and somewhat less by other chip-making machine manufacturers. Press manufacturers would not refer to the introduction of digital control as a 'technology shock'. The different qualities of a technological change are displayed in the technology shock matrix in the diagram below.

Figure 4.9: The effect of NC/ CNC on producers of different machine types

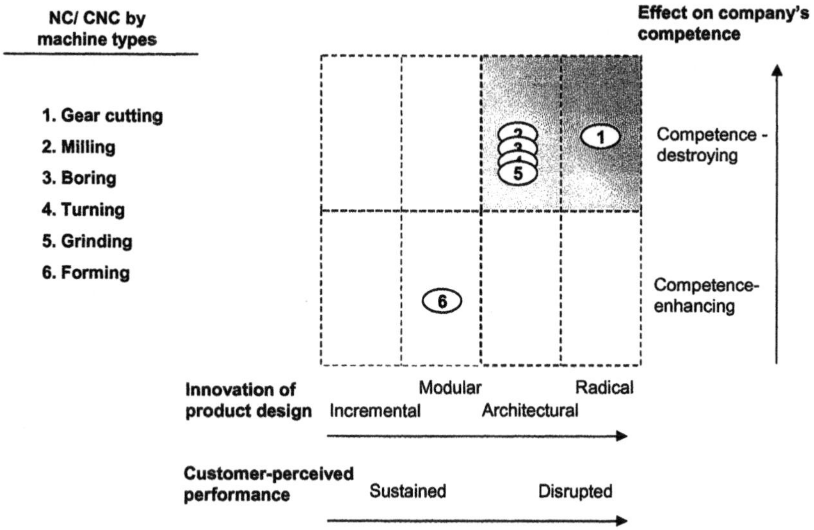

Source: own depiction.

4.6 The Japanese success story

Japanese machine tool manufacturers enjoyed market success along with the technological change. Originally, Japanese manufacturers produced machines under licensing agreements with foreign companies. For example, Mitsubishi Heavy Industries purchased the technology to produce copying lathes from the Swiss company Oerlikon, boring and milling machines from the Italian company Innocenti and gear hobbing machines from the German company Lorenz (Lee, 1996) between 1954 and 1961.

As it was described before, customer brand awareness and the perceived high level of mechanical accuracy of the established companies posed a barrier to market entry for new machine tool manufacturers. These barriers disintegrated more and more with the advent of numerical controls. Price became the dominant feature of competition. Between 1970 and 1980, the price on average for a Japanese NC machine tool was half or one-third of the price of a comparable US-produced machine (Mazzoleni, 1997). The Japanese price advantage was amplified by comparatively low wages (Unger, 1985) over a long period . In addition, the Japanese focused on NC technology and the integration of computers in the control unit at an early stage as mentioned in the section on NC's disruptive effect on company competence. In the 1970s, the percentage of numerically controlled machine tools versus the total number of machine tools was 20% in Japan, whereas the number for European and US manufacturers was 10% (see figure below)[72].

While US companies failed to benefit from their early lead in the new technology, Japanese companies followed their goal systematically. Japanese companies also pushed NC/ CNC technology, especially during the years of the two oil crises in the 1970s when other companies reduced their commitment to new technologies. The share of digitally controlled machines sold rose steadily.

Machining centers were designed as standard machines with very few options or peripheral equipment for the customer to choose from. Worldwide marketing and a small number of variants allowed for high production quantities. The diffusion of the new technology overtook US numbers in 1976 and remained at the highest level in the world from then on. Supported by a favorable exchange rate, the percentage of all machine tools worldwide produced in Japan grew from 13% in 1976 to 30% in 1980. For the European Community (EC10), which then, with the exception of German companies, was mainly competitive in the European market and even

[72] Numbers taken from AMT (1990), VDW .

lost ground there (Anonymous, 1985), that number dropped from 41% to 35%. As described in sections 4.1 and 4.6.1., this occurred at the time of the second technology wave when the US also lost its market leadership position. At the same time the number of all machine tools produced in the US fell from 38% to 24% (Harrop, 1985). The success of the Japanese companies had such a strong influence on the minds of decision-makers that some of them regard Japan as the catalyst of the market evolution since 1978[73].

Figure 4.10: Levels of NC penetration in the important producer countries

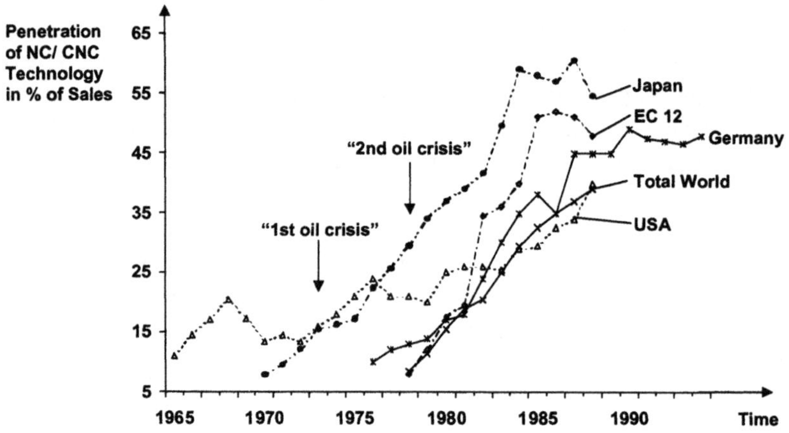

Source: own depiction based on data from VDW, Strategic Study on EC Machine Tool Sector, WS Atkins Management Consultants in assoc. with Ifo-Institut; numbers 1976, 1977, 1990, 1991 from Am. Machinist; numbers before 1979 from Roberto Mazzoleni (1997); Germany data from Schwab (1996).

Japan's entry of machine tool users into the market for machine tools was unique. With car production growing, car assemblers established their own subsidiaries or strengthened their machine divisions to produce the machines they needed[74]. In 1956, Toyoda established a subsidiary, Toyoda Koki, to produce transfer lines, and Nissan Motor cooperated with its member company Hitachi Seiki for the same purpose. In the same way,

[73] Robert Seach, Hardinge, Industry Expert Interview Series (2000-2001).

[74] An interview with Mr. Okitoma and Mr. Yoshida from Mitsubishi Heavy Industries at the EMO 2001, Industry Expert Interview Series (2000-2001), revealed that the Machine Tool Division's primary task is still to supply equipment for the special purposes of Mitsubishi Motors. The division still works as a cost center. Mitsubishi just sells machines externally if it has extra production capacity.

Mazda Motors started to make transfer machines in its machine division (Lee, 1996). This development was seen as a national effort and was largely encouraged by MITI, the Japanese Ministry for International Trade and Industry. In the early years of development, MITI provided focused resources for the transfer of technological know-how to Japan.

The successful entry of user firms is also being attributed to 'keiretsu', the conglomerate structure of Japanese industry. Risk sharing and collaboration made it less vulnerable to short-term performance pressures. Entries from various sectors increased the number of firms which were engaged in manufacturing metal-cutting machine tools from 108 in 1951 to 351 in 1959 and to 671 by 1969, an annual growth of about 10%. During the 1970s this number rose again by 6.7% (Lee, 1996). In the 1980s electronics companies entered the market and made an effort to incorporate minicomputer technology into the control of machine tools. The great success of Japanese companies is often attributed to this effort. The prominent example is Fujitsu. This company established an independent firm, Fujitsu-Fanuc, in 1972 based on its development division for computer-control devices. Its task was to market this technological asset. Fanuc became the leading producer of NC controls by the end of the 1980s, with 50% of the world market share and 80% of the market share in Japan (Lee, 1996).

Dr. Seiuemon Inaba, the guru of Japanese NC technology, also worked at Fujitsu-Fanuc. He later became president of Fanuc. He had received the plans of the NC project at MIT and had successfully completed a reproduction of such a machine in 1957. With a simplified control unit architecture, Fanuc soon came to dominate the Japanese market[75]. The Japanese production focused on narrowly-defined product offerings of relatively simple machine tools, whereas more complicated ones were traditionally imported from German, Swiss and US companies.

[75] For anecdotal evidence on Japanese control unit design and its effect on diffusion, see Mazzoleni (1997): In a joint project with Makino Milling a control unit in open-loop design which was commercialized in the mid-sixties was developed. In addition to an open-loop, the closed-loop architecture of MIT contained a feedback circuit from the work table to the command unit to increase positioning accuracy. Leaving out this rather complex element for the open-loop design promised cheaper control units. To achieve the same accuracy without a feedback circuit, the performance of the stepping motor had to be improved. In fact, the technical key parameters of stepping motors like stepping rate and torque output improved between 20 and 40 times. The closed-loop design dominated Japanese machine architecture, which focused on low costs until the early 1970s. For international markets, though, the closed-loop approach was followed again.

4.7 External factors

Unfavorable **exchange rates** hit the US machine tool industry especially hard in the 1990s. With a strong value of the US dollar as compared to the European and Japanese currencies, US manufacturers faced a cost disadvantage that they could hardly match. In the second half of the 1990s, the European currency union helped stabilize European exchange rates and put them at a comparatively low value - a development welcomed by European producers[76].

The dollar value had already decreased during the **oil crises** of the 1970s[77]. The crises started recessions that had negative effects on European machine tool exporters and hit the industry in Europe and the United States with a two to three-year delay. Demand dropped and there was overcapacity. Western manufacturers perceived it as the first tangible crisis since World War II. Cash for technological development, establishing know-how and training, etc. and even the financing of new orders became difficult. Some companies that were financially not sound enough, such as Burr and Hüller in Germany, went out of business or were bought out[78].

The **collapse of the East Block** after 1990 also meant the loss of an important export market. Although business with Eastern Europe always meant a high level of risk, the political and economical changes in the East affected German companies in particular. The economic change coincided with the worldwide economic crises, which in Germany was partially covered by increased investment activities in the former German Democratic Republic[79] following reunification. As a result of the economic crisis, cost awareness became the top priority for most manufacturing companies. Suppliers for the automotive industry experienced the 'Lopez-effect' of meeting strict cost goals[80].

The 1991 **Gulf War**, which imposed some restrictions on international trade, had a negative impact on export-oriented industries. At this time, the machine tool industry had followed the general trend of the manufacturing industry towards **globalization**. The effects of slow orders could thus be

---

76 Mr. Abler, Liebherr, Industry Expert Interview Series (2000-2001).

77 October 19, 1973 and November 8, 1978.

78 Wilhelm Gauch, Ex-Burkhardt&Weber, Dieter Hofmann, Ex-Heckler&Koch.

79 Ludwigsburger Maschinenbau GmbH Burr, who designed the first multi-spindle application, had severe financial problems after a contract with the Soviet Union was not complied with. Helmut Schneider, Ex-Burr, and Alfred Häcker, EMAG, Industry Expert Interview Series (2000-2001).

80 Dr. Lothar Ophey, Ex-ExCellO, Industry Expert Interview Series (2000-2001).

felt everywhere[81], quite a contrast to the effects at the time of the **Korean War** that had led to an increased demand in the US in the at that time regionally-structured manufacturing markets of the 1950s.

In some countries, the strong position of **labor unions** restricted the flexibility of company resource allocation. There was a great deal of discussion about overtime and vacation time in Germany. A change in workers' rights in these areas was still taboo. For financially burdened companies, personnel costs of over 40% (VDMA, 1993) without much possibility of flexible allocation to demand remained a structural disadvantage versus countries with less formalized workers' rights. In contrast, Japan's workforce has been known for its flexibility, which is still prevalent today[82].

Another factor that left its imprint on the success of medium-size, family-run businesses is the **ownership structure**[83]. Constellations where a member of the family who controls the majority of shares is at the same time managing the business seem to be prone to the risk of extreme decision-making; either radical change or no change at all.

This section demonstrated that external factors[84] and the dependence of national and world economic trends have contributed to some of the changes in the competitive landscape in the machine tool industry. None of these factors, though, overweigh the effects of technological change as discussed in previous sections of chapter 4[85]. Instead, they amplify the effects of technology shock. As a result, the basic hypothesis that technological change is the main trigger for turbulence within this market still holds true.

4.8 The next wave - trends and future technological changes

A Delphi study was carried out in order to show that the historic events observed can repeat themselves and that results obtained from these events have practical relevance (Arnold, 2002). The purpose of this study was to

---

81 Werner Babel, Ex-Maho, Dr. Udo Schapp, Gleason-Hurth, Dr. Walter Eggert, Gleason-Pfauter, Industry Expert Interview Series (2000-2001).

82 Yushiyuki Kanesige, Mori Seiki, Prof. Dr. Franz Klenger, Ex-Gildemeister, Ex-Heidenreich&Harbeck, Industry Expert Interview Series (2000-2001).

83 Frank Baumbusch, Ex-Pittler, Witzig&Frank, Industry Expert Interview Series (2000-2001).

84 The factors described here are the ones that were brought forward most frequently and named as the most important by practitioners.

85 Most of the experts agree with this conclusion. See, for example, Klaus Rietschel, Ex-Niles Chemnitz, Industry Expert Interview Series (2000-2001).

detect indicators for technological trends that can potentially hit the industry like a shock. The participants were from 7 countries: Germany (D), United States (US), Switzerland (CH), Japan (J), Italy (I), Taiwan (TW), and Spain (ES), covering 7 of the top 10 machine tool exporting countries (VDW, 1999a). 28 of the participants were from Germany. The participants predominantly held positions as managing directors or vice presidents.

Detailed results from the study[86] are not repeated here. The two most eminent results were: Technological trends which are more closely related to the company competence receive greater attention than changes which might pose more danger to the survival of a company but are not included within the scope of classical competence. When evaluated using the technology shock matrix, 4 out of 18 trends identified have the potential to become technology shocks. These are: external shocks, networked businesses, new production technologies, and new materials. New production technologies (e.g. laser cutting, electrical discharge machining, ultrasonic machining) and new materials (e.g. compound materials, plastics, coatings on tools) are rather self-explanatory. External shocks relate to potential drastic 'upstream' and 'downstream' changes which the machine tool producer cannot influence. A drastic change in products for which the machines are being utilized, could make them obsolete for a number of applications. If, for example, the combustion engine would be replaced by fuel cells or solar power, most of the complex mechanical parts of today (like camshafts) would simply disappear - and with them the machines that produced them. These shocks can potentially occur in any of the customers' industries. They can be caused by the use of new material requiring different machining, by the substitution of one module by another, etc. However, upstream shocks can radically affect the industry, too. Developments in control technology, materials, etc. are typical upstream events.

Networked business is the fourth trend which might need the most explanation. Shifts in the value chain can be expected with the connection of machines to the Internet. New business models become feasible, detracting from the value added by traditional machine tool manufacturers. A new group of 'players' could appear on the scene: contractors able to run plants without possess to have the know-how of manufacturing machines or knowledge of the final product.

Full availability of machine data can lead to diverse competitive behavior. Companies will be able to upgrade the quality of their customer relations. Warranties for equipment are usually not valid for more than a year;

---

[86] The study can be obtained from INNOtec upon request.

the reason being that with time, the chances increase that machines are not used or serviced correctly. Electronic protocols can indicate potential misuse and the data made available off-site. Together with sensor data on parts that are subject to regular wear and tear (e.g. leadways and sealing rings) and maintenance of service intervals, the manufacturers' risk of lengthier insurance periods can be reduced. Moreover, the manufacturer can verify current up time warranties of 97 to 99% by means of a system resembling an online 'trip recorder'. With the availability of lifetime data, the second-hand machines' market can operate much more precisely. Warranties can thus be extended considerably, making it possible for manufacturers to offer substantial benefits to their customers. Furthermore, machine data collected over a lifetime is valuable feedback for improvements in machine engineering design. Insurance companies, banks, and full-service leasing companies will equally value this kind of data.

Specialized global service and maintenance companies are able to establish themselves as soon as data about machine status and maintenance requirements are on the network. Online test runs of production facilities are even feasible. The trend towards further standardization will support efforts to build maintenance organizations independent of machine manufacturers. Globally-established companies, especially control manufacturers like Siemens, are testing the setup of Internet platforms. They are used to support services for machine tools offered by manufacturers or special suppliers[87].

Standardization in combination with the possibility of networking might even revive the establishment of such new business models such as e-commerce platforms. With qualified and comparable data available, products and prices become more transparent and easier to compare. Another business, that of planning services, will also experience a new quality. Customers will be able to submit their production requirements for a future plant and then take a virtual walk through their future facility, where they can have a look at a simulated production run.

[87] September 2001, Siemens A&D, Nuremberg, established an independent subsidiary, ePS Network. ePS (electronic production services) provides customers with production process data via the Internet. For this reason, the CNC/SPS controls of machine tools are linked to ePS servers. Data is analyzed and results (machine condition, process stability, etc.) are sent back to the customers. Siemens offers this service for its Sinumerik controls of the 840D series. See: http://www.ad.siemens.de/meta_sixcms/sixcms/detail.php3?template_id=218&id=2610.

The results were presented on several occasions to industry experts. The majority agreed with the results of the Delphi study.

4.9 Implications for an analysis of technology shocks in the machine tool industry

The introduction of numerical controls was much less an evolutionary process than a series of technology shocks that struck in three waves. Established companies in many cases failed to judge the full extent of change. The vanishing of leading companies from their dominant position cannot be attributed to consolidation within the industry. Due to a traditionally confusing product variety, scale effects on the production side are relatively weak - size counts when it comes to financial issues.

The turbulence in the market appeared to have occurred to a large extent due to technological change. The market segments themselves were redefined through the convergence of functionalities and the integration of the machine environment into systems. Three categories serve well to identify shocks - changes in the product architecture, changes in competence that lead to competitive advantage, and changes in the value proposition as well as the way the value of the product is perceived. The extent of threat to established businesses and the degree of opportunities for new ones depends on the product groups. For the manufacturers of presses, the threat was by far less than for the manufacturers of gear cutting machines, for example. Observing current technological developments reveals trends that can potentially have shock-like effects on manufacturers of machine tools in the future.

The industry in the UK is an example where technology was not observed carefully enough and where it was not attributed an adequate importance. This would be consistent with the view that the ability to forecast (see section 2.3.1) contributes to the success of companies.

The success of the Japanese and the failure of many US companies illustrates that focusing systematically on the order-of-magnitude economic gain for the customer is more important than the location of invention. Several manufacturers, however, did not realize that considering all potential options a customer might desire does not necessarily lead to the customer's greatest possible benefit.

The strength of the technology shock 'introduction of numerical controls' was different for the manufacturers of chip-making and forming machines. The shock was comparatively small for press manufacturers, whereas it was much stronger for the manufacturers of chip-making machines (especially for gear cutting). In a sample containing the manufacturers of presses and chip-making machines, it can be expected that factors that influence firm performance after the shock do not have the same impact. Either the relevant factors are different or their effect is much stronger for the producers of chip-making machines.

The technological change occurred in three waves, of which the last one is still ongoing. PC environments, networking capabilities and the expectation to hook up production systems to the Internet have the potential to further upset the current market structure in a way similar to the recent history of the machine tool industry.

5 Descriptive analysis

5.1 Company data and technology shock waves

5.1.1 The original set of variables

Data on a sample cohort of 58 international machine tool companies was collected from three industry-specific shock waves between 1955 and 2001. According to the technology shock matrix developed in Chapters 1 and 3, the destructive potential of the shock can be measured along two axes. The devaluation of classical competence is one axis; the change in perception of value expressed in fundamental changes of product architecture is the other. The degree of success to which companies managed technological change was measured in four ways.

The occurrence of hazards in the form of **bankruptcies, settlements and takeovers** are an indication of **failure** (Hill et al., 1980). Companies that experienced one or the other form of such hazards between 1955 and 2001 will be categorized as 'losers', companies that did not as 'winners'. As 'losers' usually experience more than one failure, an index was also created listing the number of failures of a company and thus establishing a weighted scale. The precise procedure for the construction of the scale is described in section 5.2 'The manager's mind - The perceived importance of success factors for the survival of technology shocks'. Finally, the span of time between the moment the shock occurred and the actual failures was recorded. For 'losers', the duration up to the failures was registered; for 'winners' the censored duration (i.e. the time until the end of the observation window, in this case December 2001) was noted. The duration between the occurrence of a shock and the occurrence of a failure represents another means of measuring the endogenous variable[1]. An analysis was then carried out to study the market share of machine tool companies in

[1] In a later section the method of calculating a comparable market share figure for companies from different countries will be discussed.

view of revenue. Although a rather limited measure for the success of a company, it proved to be the only financial company data that could realistically be collected for the 58 companies involved.

The companies are described by variables, which will be used to explain the success or failure of companies. These variables build on existing research - as laid out in Chapters 1 and 2 - and are supplemented by aspects suggested by academics and practitioners during previous rounds of interviews and pre-tests as described in Chapter 3. The diagram below (figure 5.1) shows the categories of variables grouped into three blocks of hypotheses: the first set contains 11 categories that describe company attributes that existed before and during the shock, the second set comprises 8 categories describing reactions to the shock in the form of strategic decisions, and the third set consists of 8 categories focusing on operational decisions and implementation programs. Each category itself comprises one or more variables. Figure 5.1 illustrates where the different issues were brought up.

Figure 5.1: Overview of variable categories

Endogeneous Variables	**Hypotheses block 1: Attributes of companies**	**Hypotheses block 2: Strategic Decisions**	**Hypotheses block 3: Operational Decisions**
• Winner/ loser • Index scale of number of failures • Duration of survival after technology shock • (Change in World market share)	• **Ability to forecast** • **Resistance to innovation** • **Financial base** • Information flow • Company culture • In-house manufacturing • Attractivity for new hires • Management focus • Flexibility of plan • *Distribution of power* • *Consequence of decision making*	• **Innotype 1** • **Innotype 2** • **Timing of market entry** • Organizational independence • Acquisitions • Alliances & JV • Changing the value chain • Business model	• **Resource commitment** • **Cooperation with suppliers and customers** • **Trainings** • **External R&D** • Level of hierarchy • Extent of implementation program • New hires • *Extent of decisions*

Legend: **Bold** - factors which had significant results in previous research on technological change.
Italics - factors suggested by practitioners.
Source: own depiction.

Typically, attributes of the type studied here develop over a long period of time within a company and cannot be changed rapidly. They are already in place prior to a technology shock. Several company attributes, however, are strongly linked to the personalities of some of the organization's key people: particularly openness or resistance to innovation, focus by management, company culture, consequence of decision-making, and distribution of power. On the other hand, the ability to forecast, planning flexibility, information flow, financial base, the know-how of manufacturing in-house, etc. need time to develop. Thus company attributes can be observed even before a technology triggers a shock wave. However, they are still applicable when the shock occurs - most likely even characterizing the companies after the shock.

The second set of hypotheses is operationalized through variables that emphasize strategic decisions that have to be considered in the face of radical technological change. They can only be made once a new technology already exists.

The third set emphasizes operational aspects of resource commitment, development of know-how, and other implementation-related aspects. Similar to the strategic decisions, these aspects only become relevant and are decided upon after the shock.

Figure 5.2: Time relatedness of variables

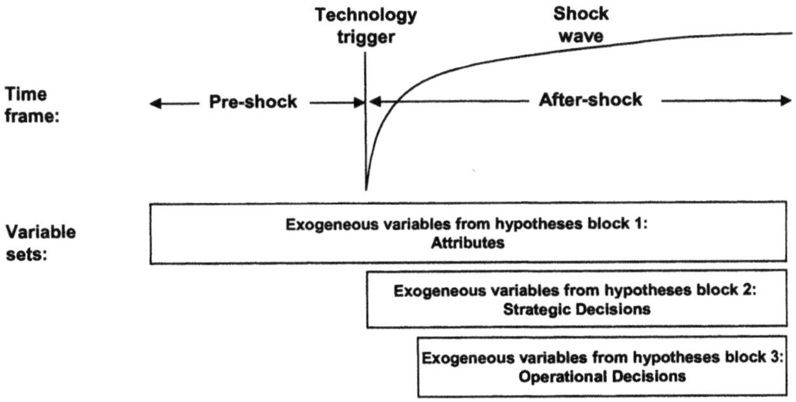

Source: own depiction.

Tables A.5.1 - A.5.3 in this chapter's appendix explain the three sets of hypotheses. They list the variables according to their categories, the questions used in the questionnaire, the variable shortcuts, the type of operationalization, and the range of measured values. The variables' characteristics are described in tables A.5.4 - A.5.6 of the appendix.

5.1.2 Companies involved and coverage

The 58 companies involved cover a fairly large proportion of the world market. Their accumulated sales represents between 12 and 18% of the total world machine tool market for the years indicated. 56 of the companies produce metal-removing machines. Their accumulated sales represents between 17 and 24% of the world market share in the metal-removing machine tool industry for the years indicated.

Table 5.1: World market of machine tools and share covered by sample

Year	1960	1970	1980	1990	2000
World market, all machines, in Mio US$	3700	8000	26800	45300	30600
Share covered by sample	12%	16%	18%	18%	16%
World market, metal-removing , in Mio US$	2500	5500	18500	31700	21100
Share covered by sample	17%	22%	24%	24%	20%

Source: own aggregation of news releases, publications from industry and trade associations.

In the year 1970 - prior to the drastic technological change due to through the introduction of CNC - 25% of the companies in the sample engaged less than 200 employees, 25% between 200 and 500 employees, another 25% between 500 and 1,000 employees, and the last 25% more than 1,000 employees. In a sense, the sample is fairly representative for the machine tool industry, which, particularly in Germany, is characterized by mid-sized companies, and where 50% of the companies in the sample are located. Table 5.2 provides an overview of the development of percentiles of size in the sample over time. The distribution of size did not change dramatically over the years.

Table 5.2: Percentiles of numbers of company employees

Year	1960	1970	1980	1990	2000
25 percentile	100	200	200	318	240
50 percentile	375	500	560	600	400
75 percentile	838	1000	1500	1200	1050
Maximum	10000	12000	12000	7416	10000

The sample results from a list of 276 worldwide companies identified as the most important[2] firms and thus approached for participation in the study. Companies that could be located were asked to participate. Table 5.3 gives an overview of the companies' headquarters as well as the country-specific response rates. Most of the participants are from Central Europe. The response rate from the Taiwanese contacts was relatively high. German companies are strongly represented in the sample; 50% of the participants are German. With 32%, German companies are thus strongly represented in the list of the 276 most important companies worldwide. The highest participation came from companies from the former East Germany and Switzerland.

Table 5.3: Origin of the most important companies and their response rates

Country of origin	Number of companies	Of which participated	Response rate
Austria	2	-	0
Switzerland (CH)	21	9	43%
(West) Germany (D)	86	29 (+1 after merger[3])	35%
East Germany (DDR)	6	4	67%
Spain (ES)	10	1	10%
France (F)	16	1	6%
Italy (I)	27	4	15%
Japan (J)	26	4	15%
Korea	4	-	0
Netherlands	1	-	0
Taiwan (TW)	12	3	25%
United Kingdom	10	-	0
United States (US)	45	3 (+1 after merger[3])	7%
Origin unknown (xy)	11	-	-

The 58 companies that actually participated represent 21 % of the total of 276 companies approached. A complete listing of the 276 important companies can be found in table A.5.7 in the appendix to this chapter.

---

2 Selected on the basis of statements from interviews with industrial experts, consulting reports, universities, research articles with a historic perspective, newspaper articles, printed company advertising material and histories, company histories on the Internet, the AMT Blue Bulletin, professional databases, and industrial association publications.

3 The company carried a different name after its merger with another firm.

5.1.3 Differentiating success and failure

Survival, i.e. the existence of a company or division after a technology shock, is obviously the key prerequisite for its success. Nevertheless, survival is a comparatively weak measure since it does not differentiate between the various survivors. Other commonly-used indicators are market share and citations from business literature like annual reports, business articles, industry surveys, stock reports, etc. (Cooper et al., 1992). Another indicator is the development of an organization's value (expressed, for example, in stock and bond ratings) relative to its competitive environment (Smith et al., 1988).

Most of the literature examined deals with relative failures and attempts to distill recommendations for avoiding resulting failures (Christensen, 1997; Christensen et al., 1996; Levitt, 1960; Myers et al., 1978). In Christensen's and Suarez' study the measure of success was a company's ability to stay in the new market or its respective exit. As already mentioned above, this test lacks differentiation and only offers conclusions on how to avoid the fatal extreme of market exit. It does not distinguish good performers in a market from less successful ones that nevertheless remain in the market.

Many studies[4] use the judgment of involved management to differentiate success and failure. Success judged by the people involved is prone to personal bias.

Willard and Cooper investigated differences between success and failure for industry shakeouts in the US color television set industry. They identified a number of success factors but did not relate them to radical technological change (Willard et al., 1985). Although this approach includes elements that explain survival in the face of technological change, an observation period over the entire lifetime of the industry includes various effects (e.g. overall consumption, stock market fluctuations). As displayed in figure 5.1, this research makes use of four variants to assess success and failure:

- The dichotomous differentiation between winners and losers is used for the descriptive analysis in this chapter.
- An index scale constructed from the frequency of the companies' failures is used in section 5.5.6 and for multivariate analysis in chapter 6.
- Information on the world market shares for each of the companies in the sample has been collected for the years 1960, 1970, 1980, 1990, and 2000 (see appendix A.5). As a tribute to the readability and volume of

[4] e.g. Ebadi and Utterback (1984).

this thesis, the information is not used to form another endogeneous variable. It is used to describe the sample as in section 5.1.2 and table 5.1 and for the imputation of missing data in chapter 6.
- The duration of survival after a technology shock is used for multivariate analysis in chapter 6.

5.1.4 The 'Zombie' phenomenon - or why counting market exits does not always work

In this sample of the 58 companies, 70 hazards (bankruptcies, settlements, -or takeover) occurred. 63 of them (48 takeover and 15 bankruptcies) would usually have resulted in the disappearance of companies from the market - to put it bluntly, it would have been more than enough to have all of the companies disappear from the market. In addition, 7 settlements with banks or suppliers took place. In reality, the 70 hazards occurred in only 32 companies; 26 firms never experienced any failure throughout the observation period. Out of the 32 companies that failed, only 8 disappeared from the market. Thus, counting market exits as a measure of change or activity in the industry - as is done in faster-moving industries (Christensen et al., 1996; Klepper, 1996) - is not a viable method in this case. It would produce the wrong results. As for the machine tool industry, using the total number of firms in the SIC group does not seem to work either and counting entries and exits based on industrial reports is also no viable alternative (moreover, it is not even available in a usable form). Several ownership transitions and bankruptcy filings happened within one and the same company during its history.

14 of the 32 companies that failed went through this experience only once. The other 18 companies failed twice (8 companies) or more often (10 companies). Many companies filing for bankruptcy continue operations with fresh money from either their own investors or other investors after the bankruptcy. Real shakeouts are rare in the industry[5]. Companies are either revived by owners' families, banks, investor consortia, investments by management or employees, or by competitors who acquire equity shares (usually only for a certain period of time) and subsidize operations of the affected companies. When these companies are taken over, they still

[5] The statistical department of VDW (Mr. Geis, 6.11.2001), Frankfurt, shares this observation. 'Real 'Marktbereinigung' (cleansing of the market) has not happened in the industry - this is what companies tell us. Numerous companies still exist after MBOs, additional support from banks, groups, or are just 'the living dead'. However, there is a trend towards group structures (Thyssen Production Systems, IWKA, Gildemeister, UNOVA).'

remain independent legal entities. Investors' motivations to become involved differ. Analysis of the companies' histories revealed several motivational factors. For some it was the social commitment to secure employment, for others the potential opportunity to grow in size, for others again it was taking on political influence, and finally for investors the sunk cost phenomenon - the fear of losing their previous investments. There are, of course, more factors involved and in most cases it is a combination of the above that allows failed companies to reappear on the market after a hazard. Common in many instances where a company that had failed was reanimated was the expectation that due to a seeming inertia in the market (slow pace of technological advance - i.e. rather evolutionary, low customer churn), the company would have a good chance of catching up with its competitors.

The negative 'champion' in the sample is a company that experienced 8 failures during the observation period. Two companies experienced 5 failures, one company 4 failures, and 6 companies 3 failures.

5.1.5 Defining the shock waves and expanding on the characteristics of the 4-stage wave pattern

Technology shocks trigger waves of competitive change which are, according to the theory that was elaborated on in chapter 1, characterized by four phases. These four phases include the introductory state, the rapid growth phase, the establishment of a dominant design, and, lastly, cutthroat competition. The phases unfold after a technology shock has punctuated a previously existing quasi equilibrium. The data series are grouped into 'winners' and 'losers' according to the 'failure' criterion (i.e. winners - companies without bankruptcy, settlement, and not having been taken over; losers - with at least one bankruptcy, settlement, or having been taken over). External shocks punctuating the equilibrium occurred in three waves as described in the Chapter 4 discussion on the history of the machine tool industry: the core period of the first wave was between 1955 and 1965, the second one from 1974 to 1984, and the third one began in 1990.

For the purpose of this investigation, the first introduction of a product on the market is regarded as the trigger or technology shock. Thus, the **first shock** caused by digital control technology was the one caused by the introduction of the first commercial NC machine by Giddings&Lewis (Romeo, 1975) in 1955. CNC machines, whose controls were microcomputer-based, triggered the second shock. Japanese firms introduced the latter type of machines to the market in 1974 (Lee, 1996). The effects of this **second shock** wave (starting in 1974) were amplified by a massive world market entry by Japanese companies. The **third shock** was caused by PC-based controls. Products with this technology appeared on the market around 1990 (Chalmers, 1998). The three technological changes that triggered the various shock waves are displayed in figure 5.3 below.

Although small in size, the sample reflects a wave pattern typical of technology diffusion. Failures in the third wave are very much evident in the sample. The increase in failures is congruent with what was defined above as the beginning of the third wave in 1990. The second and third waves appear as two smaller 'bumps', called 'early' and 'late' waves. The second 'early' wave begins after the first oil crisis in the second half of the 1970s and lasts until around 1980. The second 'late' wave appears next to it and seems to ebb off in the mid-1980s. The beginning of it coincides with the massive market entry by Japanese companies. Firms that experienced failure during the second 'late' wave were mainly manufacturers of milling machines and lathes. Substitution occurred above all in these segments. The first digital machines, which were largely successful, were machining and turning centers that substituted the classical milling machines and lathes.

Figure 5.3: Failures in the sample of 59 companies and shock waves triggered by numerical controls

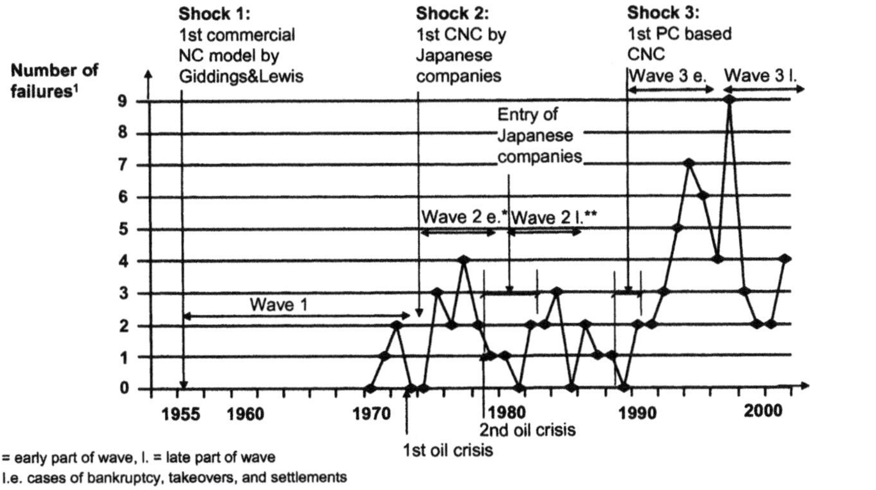

e. = early part of wave, l. = late part of wave
* I.e. cases of bankruptcy, takeovers, and settlements
** esp. Grinding machines (also milling, lathes)
*** esp. Milling and lathes (subst. by machining centers and turning centers)
Source: own depiction based on sample of 59 machine tool manufacturers.

The question now arises of how to relate failures and shocks in the sample. Although the final triggers for failure were frequently trouble with owners, some major write-offs from failed contracts or settlements with creditors, the technology waves appear to coincide with the relative maxima of the failure curve. The failures during the last wave occurred at the same time as the recession and, according to industrial experts, were mainly triggered by crushing markets yet intensified by the technological change.

The hazards that occurred between 1970 and 1974 should clearly be attributed to the first wave. After the second shock in 1974, one might consider an overlap of shock waves. The same situation could have occurred after the third shock in 1990 - the effects of the previous waves still being felt while the next shock was already taking place.

According to the theory developed in chapter 1, it would take several years for the effects of a shock wave to turn into hazards during the phase of cut-throat competition. However, there were not only 'late' failures (waves 2l and 3l) but also 'early' failures (waves 2e and 3e).

As for this investigation, this might mean that relating hazards from 1975 to 1989 to shock 2 would not be accurate. Some of the failures (especially failures of wave 2e) could have been caused by other triggers, or even shock 1. The same applies to failures between 1990 and 2001: some of the failures (especially the early ones up to 1996) could have been caused by triggers other than those of technology shocks, or even be a result of shock 2.

Different approaches of dealing with this situation could be considered:

1. Suppress early failures of wave 2e and those between 1990 and 1996 in the analysis as 'not being related to technology shocks' and regard these companies as 'winners'.
2. Regard failures in waves 2e and 3e as results of the proceeding shocks, i.e. shocks 1 and 2.
3. Focus particularly on the analysis investigating the differences between 'early' and 'late' losers. In this way, it might be possible to find indicators that could help expand on the existing theory and explain the failure mechanisms of waves 2e and 3e as a result of the respective shocks 2 and 3, even though they occurred shortly after the shocks.

The first two alternatives would seem to be rather random rules regarding the treatment of data and superimpose a certain expectation of resulting failure patterns. For this reason, the third alternative was chosen. The analysis will be geared towards the completion of characteristics describing the 4-phase wave pattern introduced in section 1.3.

5.1.6 Number of firms under observation

This sample contains datasets from a total of 58 firms between 1955 and now. 55 firms were under observation for this research during the second wave. 13 of these firms experienced failures during the second wave, 3 firms prior to that in 1971 and 1972. The latter failures could be the result of some late effects of the first wave. For statistical reasons, these 3 firms together with the other 13 form one group of 'losers' totaling 16 companies dating back to the two early shock waves. 25 of 53 companies experienced failures during the third shock wave starting in 1990. 25 companies did not experience any failure at all during the observation period. These companies are called the overall 'winners'. Expanding on the existing theory, one could try to explain the mechanisms of early failures during waves 2e and 3e as a possible result of the respective shocks nos. 2 and 3, even though they occurred shortly thereafter.

Table 5.4: Winners and losers frequency during the shock waves

	Companies total	Winners	Losers		
			Total	Early	Late
Shock wave 1	49	46	3	-	3
Shock wave 2	55	42	13	6	7
Shock wave 3	53	28	25	14	11
Total number of different comps.	58	25	33	-	-

Thus, three types of comparisons between individual samples are possible:

1. Application of the 'strict' criterion: differentiation between companies that never experienced failure and those that failed once or more during the entire observation period of 1955 - 2001.
2. Differentiation between 'winners' and 'losers' of waves 1 and 2 (i.e. events between 1970 and 1989).
3. Differentiation between 'winners' and 'losers' of the recent wave 3 (i.e. events between 1990 and 2001).

These comparisons are found in the following section 5.5.

In addition, differentiation between 'early' and 'late' failures within shock waves will be introduced. This sample does not include any 'early' failures for the first shock wave. As for the second shock wave, failures between 1974 and 1980 are called 'early', and the ones between 1980 and 1989, are called 'late'. Thus, there are 6 'early' and 7 'late' failures for the second wave. At the same time, the third shock wave failures that occurred between 1990 and 1996 are distinguished from those that occurred between 1997 and 2001. 14 of them are, therefore, 'early' failures and 10 of them 'late' failures.

5.1.7 The importance of the second shock wave

In many cases, the introduction of NC and CNC occurred without much delay. This is the reason why the question of second generation innovation is formulated in a rather general manner, i.e., 'What was the extent of technological change between the introduction of the first NC machine tool and successive NC/CNC machine generations' It takes into account that the machine generation following the first NC generation could have already been based on CNC technology, thus 'leap-frogging' the subsequent NC generation. As described in chapter 4, the introduction of the NC did by no means reach as many manufacturers as did the introduction of the CNC. The example of German manufacturer Hermle[6] illustrates well how close the first generation of NC machines and the introduction of CNC machines can be on a time axis. The first NC machine was ready for the market in 1975; the first CNC machine in 1978. In general, the first commercial NC machine had already been on the market in 1955; the first commercial CNC machine only in 1974. The success of the first CNC machines might have finally triggered manufacturers, who had not yet dealt with numerical control technology, to go the way of digital controls. It therefore seems reasonable that the three events of wave 1 are combined with those of wave 2 for statistical purposes - as proposed in section 5.1.6 of this chapter. Hermle's company history, in many aspects typical for the industry, is summarized in table 5.5 below.

[6] Excerpts from the company's webpage under http://www.hermle.de.

Table 5.5: 'Hermle's history at a glance'

1938	Berthold Hermle founds the company, a bolt and machine screw manufacturer in the southwestern German town of Gosheim. Production of turned parts.
1956	Beginning of production of centrifuges.
1957	The first milling machines.
1972	Presentation of the first Hermle Universal Milling Machine.
1975	Production of NC-controlled milling machines.
1978	Beginning of production of CNC-controlled milling machines.
1990	Maschinenfabrik Berthold Hermle Aktiengesellschaft becomes a publicly traded company.
1992	Start of a major restructuring of the Hermle Group.
1993	Cooperation with Traub AG, Reichenbach/Fils regarding foreign sales.
1994	1994 Traub AG, Reichenbach/Fils, acquires 50.1% of Maschinenfabrik Berthold Hermle AG's ordinary shares.
1995	Development of vertical machining centers launched on the market in 1996. Optimization of the company as part of the 'Concept 2000' program.
1996	The company is awarded DIN EN ISO 9001 certification. Private investors and company employees buy back shares held by Traub AG.
1997	Hermle + Partner Vertriebs GmbH commences operations. At this point Hermle has its own sales structure for the first time. The introduction of the U-series by Hermle sets new standards in universal milling.
1998	The turned parts sector becomes Präzisionstechnik Mahlstetten GmbH (PTM). At the end of the year PTM is sold, the focus being on the main sector, the machine tools. The HLS Hermle-Leibinger Systemtechnik GmbH in Tuttlingen is founded as a joint venture. The main task of this company is the customer-specific automation of Hermle machines and equipment with additional technology.
1999	The Hermle Schweiz AG is founded as a distribution center in Switzerland.
2000	Founding of Hermle Maschinenbau GmbH in Mössingen. This company aims at participating in the development and mounting of new machines. Inauguration of the new technology and training center at Gosheim headquarters.
2001	New production and assembly facilities at Gosheim. The compact machining center C 500 is presented at the most successful in-house fair in the company's history. The installation of a large size machining center completes the flexible manufacturing system.

The success factors as they have been asked of participants in the questionnaire are focused on the second shock wave. The questions regarding the financial base, expressed by parameters RDABS, RDPERC, RENDPERC, RENDABS, and SALES75 refer explicitly to the period of the second shock wave.

The general context given to the participants was that all questions related to the introduction of NC/ CNC (see appendix tables A.5.1 through

A.5.3 and screenshot figure A.3.1). According to chapter 4, this introduction happened in most machine tool companies between 1970 and 1990. So the answers to the questions are related to this time period. There are special cases where CNC technology was introduced for the first time parallel to PC-technology after 1990. In general, PC controls pose a further technology shock (shock 3) to the industry (see section 4.8).

In the following analysis of section 5.4 and chapter 6, the performances of companies influenced by shock 3 are put in relation to the management decisions and company attributes of the previous second wave. By doing this, the long-term effect of managerial responses on company success can be shown even after the next shock has occurred in the same technology field. Thus a total set of recommendations consists of factors that influence the short - and medium term - performance of companies (wave 2) and factors that influence the success of companies in the long term (wave 3).

5.2 The manager's mind - the perceived importance of success factors for the survival of technology shocks

The managerial view is used to achieve two goals. First, the base model to be tested in chapter 6 can be refined. It is expected that there are factors for successfully managing companies through technology shocks which are perceived as more important than the 9 factors that are - based on chapter 2 - already defined for it. These additional factors would be considered for including them in the base model.

Secondly, the priorities of management can be compared with the outcome of the quantitative analysis. Great differences between quantitative results and management's ratings can indicate a potential need to redirect managers' attention.

5.2.1 Importance ratings of success factors

Decision-makers from 58 international companies were asked to rate the importance of the 30 categories of variables related to the successful management of technology shocks. They were supposed to rate the importance of each single factor on a scale from 1 (low) to 5 (high). This process was carried out in combination with a 10-page quantitative survey[7]. The participants were informed that the questions are all connected to the transi-

[7] An account of all the questions applied is given in the appendix of chapter 5.

tion from mechanical machine control to numerical controls in their industry. They were to rate the importance of the respective factors on the success/ failure of their companies in the face of this technology shock. The types of success factors are the ones that were constructed in chapters 2 and 3: first, attributes of the companies; second, strategic decisions; and third, operational decisions. An overview of the factors included and the questions used is given in the tables A.5.8 - A.5.10 in the appendix. For each of the potential success factors, the question text is listed as well as the measurable shortcut, and the number of observations (answers) received for each factor.

In sections 5.2.2 to 5.2.4, the ratings are ranked by the relative importance which has been attributed to it by the managers.

5.2.2 Ranking by importance of companies' attributes

The chart below displays the rankings by importance of the companies' attributes.

Figure 5.4: Perceived importance of attributes

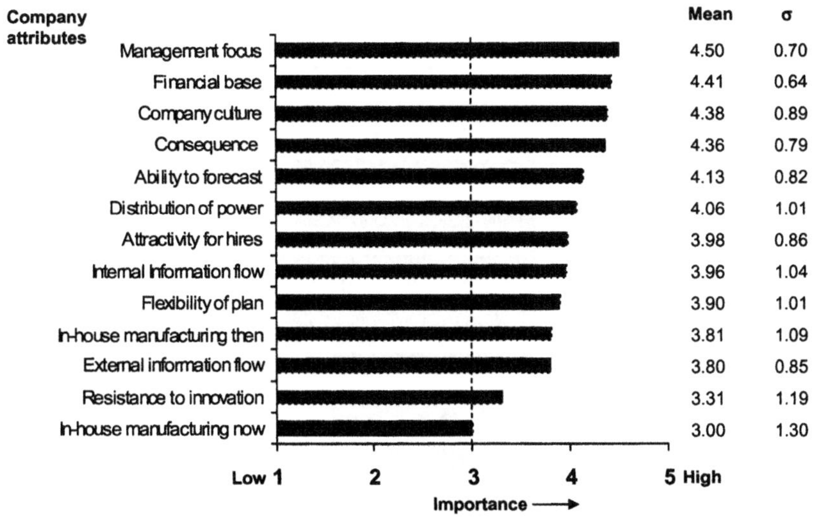

Source: company attributes, rated according to importance by decision makers of 58 companies.

There are altogether 13 factors, for which statements have been collected. They are ranked by their mean value. All but one of the attributes reach mean values above 3 - the middle value on the scale. Only the importance for a company to have the capabilities of producing in-house now

('In-house manufacturing now') was seen as medium important on average. All the other factors reached values between 3.31 and 4.5.

The highest valuation was for management focus, i.e. the priority, which technology management receives in board decisions, and the degree to which top management is involved in board decisions. The financial base, i.e. the profitability of the company itself and the resources, which have been allocated to R&D, received very high ratings. At the same time, company culture also appeared to be very important. Company culture was explained as the degree to which personal initiative was restricted or rewarded, the extent to which top management actively collects new ideas, and how good access to top management was. Timely decision-making and consequent implementation ('Consequence') was held to be very important with 4.36 on average. The ability to forecast status and trends in new technology and the correct valuation of the new market is summarized as 'Ability to forecast' and reaches an importance ranking of 4.13.

5.2.3 Ranking by importance of strategic decisions

7 ratings were expressed on the importance of strategic decisions (see chart below).

Figure 5.5: Perceived importance of strategic decisions

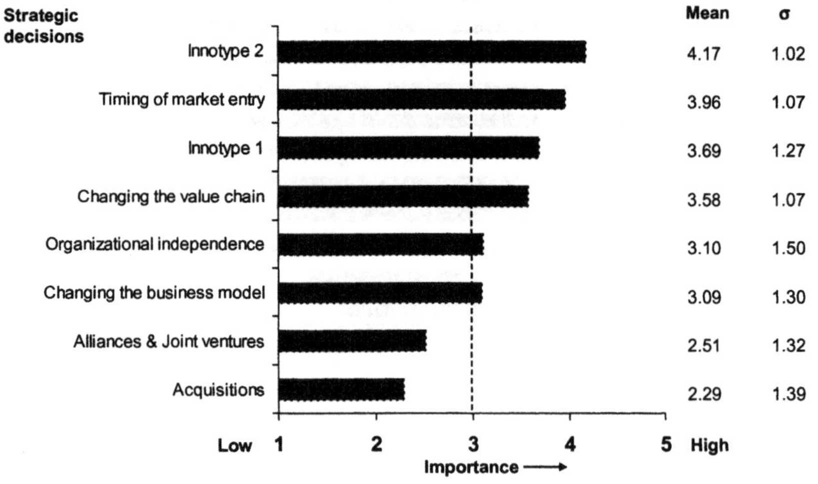

Source: importance of strategic decisions, rated according to importance by decision makers of 58 companies.

Only one reached an average value above 4 - Innotype 2 (i.e. the type of innovation in the second machine generation that followed the technology shock: incremental, modular, architectural, or radical. Innotype 2 refers to the type of innovation in the first machine generation after the shock).

Organizational independence and changes in the business model appear to be of medium importance. Alliances, joint ventures, and acquisitions are considerably below the middle of the scale. Their importance is rated lower than the other factors.

5.2.4 Ranking by importance of operational decisions

10 factors from the group of operative decisions have been evaluated. All of them are on average rated above the middle of the scale. The highest values have been given to the scope of decision-making (extent of decisions), which expresses the importance of implementation programs as being radical or incremental in nature. Relationships with customers and suppliers are equally seen to influence the success of a company when technology shocks occur. The quality of relationships includes the influence of the collaboration on the new NC - product and also the extent to which the customers/ suppliers controlled the value chain.

Figure 5.6: Perceived importance of operational decisions

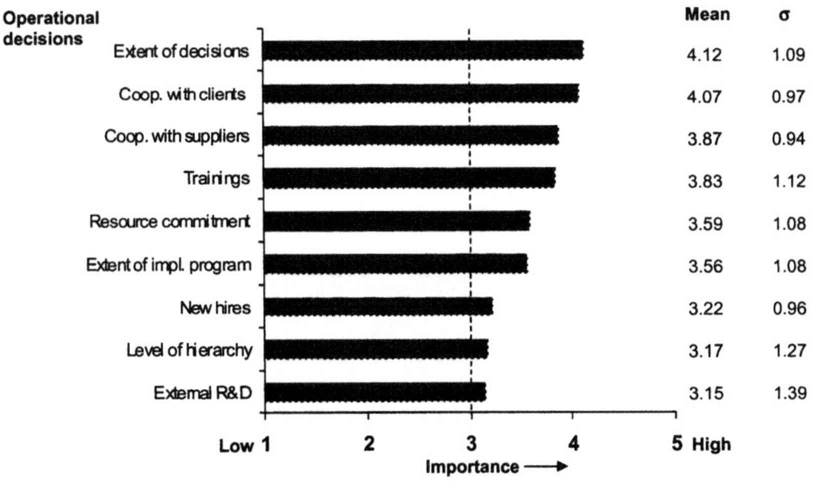

Source: importance of operational decisions, rated according to importance by decision makers of 58 companies.

5.2.5 Overall ranking with suggestion of special focus

In this section, all three groups of success factors are included in one ranking. The 10 highest rankings have averages between 3.96 and 4.5. In the list below, these ten factors are displayed according to their average importance and an indication to which group of factors they belong.

There is a dominance of factors, which are related to companies' attributes. Only one factor, 'Innotype 2' is from the group of strategic decisions; 2 are from the group of operational decisions. For future analyses, it makes sense to test whether the high importance ratings of the 10 factors hold true to statistical testing.

4 of the 9 factors defined for the base model in chapter 2 were rated as very important by the managers. Management focus is rated higher than all factors - company culture and consequence of decision-making higher than all but one of the factors of the base model. Ability to forecast and extent of decisions are close in their ratings, as are cooperation with clients and distribution of power.

Table 5.6: Factors rated highest of all by management

Factor by importance (top = highest mean value)	Variable group
Management focus	Attributes
Financial base	Attributes
Company culture	Attributes
Consequence	Attributes
Innotype 2	Strategic decision
Ability to forecast	Attributes
Extent of decisions	Operational decision
Cooperation with clients	Operational decision
Distribution of power	Attributes
Attractivity for hires	Attributes

Bold type: factors which have been defined for base model in chapter 2.

Similarly, the lowest rated factors are listed in table 5.7. They received average importance ratings between 2.28 and 3.56. An issue for future quantitative testing would be if this perceived low importance of the 10 factors below holds.

The low ranks are dominated with factors from the strategic and operational decisions group. Only 2 of the last 10 are from the attributes group. 4 factors each are strategic and operational decisions. There appears to be a great difference in the importance of the factors from different groups.

Most important seem to be the companies' attributes, and least important strategic decisions that can be made in the short-term.

2 of the factors defined for the base model are found among the lowest rated factors[8].

The predominance of factors from the 'attributes' group and the relative weakness of factors from the strategic decisions group should be further tested. If it is found that a company's attributes predominantly influence its survival of technology shocks, then there is not much that management can do to react to technology shocks on the strategic and operative levels. The decisions can be made short-term at the time of technological change, the nature of attributes, however, do not allow them to be changed quickly. Management focus, culture, consequent decision-making, forecasting ability, etc. have to be 'taught' rather than decided on. They develop inside of companies over long periods of time and are in many cases related to the personalities of key employees. The survival of companies in the face of technological change would be determined already before the change occurs.

Table 5.7: Factors rated lowest of all by management

Factor by importance (bottom = lowest mean value)	Variable group
Extent of implementation program	Operational decision
Resistance to innovation	Attributes
New hires	Operational decision
Level of hierarchy	Operational decision
External R&D	Operational decision
Organizational independence	Strategic decision
Change of business model	Strategic decision
In-house manufacturing now	Attributes
Alliances & Joint ventures	Strategic decision
Acquisitions	Strategic decision

Bold type: factors which were defined for base model in chapter 2.

[8] It might be an indicator that ratings of managers can be poor or that the factors chosen in past research are the most relevant ones to explain firm performance.

5.3 Secondary findings from the analysis of importance ratings

The ratings were collected from decision-makers in firms with significantly different backgrounds. They convey different characteristics such as the success of their companies in the face of technological change, nationality, and size. The following tests the extent to which these differences in managers' backgrounds result in different ratings of importance for success factors.

5.3.1 Differences between winners and losers

The decision-makers were all from the same industry and were asked about the same events in the past. One could expect that in retrospect, managers from failed companies had learned from the mistakes made in their companies, and that managers from successful companies had internalized what was important for their success. To test this assumption, the statements were collected into two groups: one for successful companies - the winners - and one for companies that failed - the losers. A t-test[9] is used to identify significant differences in how the importance of success factors was rated. It appears that the ex-post perspective of the participants does not guarantee the same priorities for the different factors. Below are the results for the factors with significant differences.

Table 5.8: Differences in the ratings of importance caused by success/failure

Variable	Losers		Winners		Difference
	n1	Mean1	n2	Mean2	of means
INNO1IMP	30	3.30	24	4.17	0.87**
MARKTIMP	24	4.25	24	3.67	-0.58*
RESISTIM	29	3.66	22	2.86	-0.79**
SUPPLIMP	31	4.10	22	3.55	-0.55**

Significance levels of t-test for differences of mean values: *** <= 0.01, ** <= 0.05, * <= 0.1.

Of 30 potential factors rated for importance, four showed significant differences in how they are perceived by managers of successful and failing companies. Three of these - the importance of market entry timing, the im-

[9] The sample groups were tested for normal distribution by the Kolmogorov-Smirnov test. For the parameters in table 5.8, normal distribution cannot be rejected. Thus the t-test can be applied.

portance of resistance to innovation, and the importance of the relationships with suppliers - are rated higher by the managers of companies that failed than by their counterparts in the more successful organizations. This could indicate that a lack in focus on these factors might have caused failure for the 'losers', and as a result managers who experienced failure are especially aware of them.

On the other hand, 'winners' rate the importance of the first type of innovation higher than the 'losers'. This could mean that this factor is valued highly by the 'winners' because it contributed significantly to their actual success.

Another explanation for the significantly different ratings in importance would devaluate the initial assumption that the ex-post collection of statements leads to comparable results in the 'winners' and 'losers' groups. It could be the case that the decision-makers did not reflect on the past events in their companies and that their statements reflect typical attitudes from companies prone to failure or, as the case may be, success. In other words, 'losers' would overrate the importance of market entry timing, scope of decisions, resistance to innovation, and supplier relations; and they would underestimate the importance of the first type of innovation.

To summarize, one can say that the lessons learned from success and failure are not the same for some potential success factors. In order to better interpret the differences in ratings, however, data about the extent to which the success factors are relevant for the companies has to be collected.

5.3.2 Specific characteristics of the German firms - the largest national group of companies in the sample

Over half (i.e. 31) of the 58 companies are from Germany, so it is quite possible that there is something characteristic in the attitude of German decision-makers, which could influence the overall outcome of the analysis. For this reason, the data is split into a German group and a non-German group and is tested with the t-test for samples where the assumption of normal distribution could not be rejected. In cases where normal distribution could not be assumed, the Kolmogorov-Smirnov Z-test was used. The following two significant differences were found. In both cases, the average importance expressed by German managers was lower.

Table 5.9: Differences in the ratings of importance caused by nationality

Variable	Non-German		German		Difference of means
	n1	Mean1	n2	Mean2	
INNO2IMP K-S Z-test	24	4.50	28	3.89	-0.61**
REALIMP K-S Z-test	25	3.92	29	3.24	-0.68**

Significance levels of test for differences of mean values: *** <= 0.01, ** <= 0.05, * <= 0.1.
K-S Z-test Variables tested by Kolmogorov-Smirnov Z-test for case of no normal distribution (significance levels of t-tests in brackets).

The following assumptions can be made based on tables 5.8, 5.9 and 5.10. Although the distribution of winners and losers in the German and non-German groups is not completely balanced (see table 5.10), the significant differences in the ratings are different from the ones for winners and losers. The fact that non-German managers have a higher estimation of the importance of the type of innovation on which the second generation of machines had been based (INNO2IMP) and the type of implementation program (REALIMP) than German managers seems to be independent of success and failure and is most likely related to nationality.

Table 5.10: Frequencies of winners and losers by nationality

	Non-German	German
Winners	15	11
Losers	12	20

In addition, a 'German-speaking' group was formed by grouping the 9 Swiss companies with the German companies to test the influence of a similar management culture. There is only one significant factor when the values of Swiss companies are put into one group with the German companies. This factor is different from the ones detected for the German - non-German comparison.

Table 5.11: Differences in the ratings of importance caused by language/ culture

Variable	Non-German speaking		German speaking		Difference of means
	n1	Mean1	n2	Mean2	
MANFNIMP	16	2.44	36	3.25	0.81**

Significance levels of test for differences of mean values: *** <= 0.01, ** <= 0.05, * <= 0.1.

From the 9 Swiss companies included, 3 would be categorized as 'winners' and 6 as 'losers'. Decision-makers from German-speaking countries rate the importance of having manufacturing in-house today (MANFNIMP) - although both at a low level - significantly higher. This finding implies that a conclusion about managers' mind-sets according to language regions would be misleading. If there is something like a typical national manager's mind, then Swiss is not necessarily German.

5.3.3 Differences depending on the size of companies

The perceived high importance of companies' attributes for their relative success or failure suggests testing for company size. Management behavior and focus, company culture, etc. will appear quite different for a small company than for a large one. The shock wave of introducing CNC happened after 1974. In order to obtain categories of firm sizes, the employee numbers of 1970 are used, which could be collected for almost all companies. A company smaller than 100 employees would be called small, a company between 100 and 1,000 medium, and a company with more than 1,000 employees would be called large. According to these criteria, 8 companies are small, 26 medium, and 21 companies large. 3 companies did not yet exist in 1970. A single factor ANOVA is used to test if the means of the factors show significant differences between these groups (for the case that the normal distribution assumption cannot be rejected[10]). The only factor where the means differentiate at a significance level of $p <= 0.1$ is the amount of employees devoted to the development of NC/CNC (ENGIMP). For this factor, the decision-makers from companies with different sizes attributed different levels of importance. With an average of 3.86, decision-makers from small companies express the highest level of importance. For medium companies this value is at 3.82 and for large companies at 3.06. A possible explanation could be that resource allocation for developing NC machines posed greater difficulties for smaller companies than for larger ones. In companies with more than 1,000 employees, it was obviously not a high priority to gather the employee resources necessary for development.

---

[10] The factors INNO2IMP, MARKTIMP, DECISIMP, CULTIMP, FOCUSIMP, RDFINIMP, CLIENIMP, CHAINIMP, for which normal distribution could not be assumed, have been tested using the Kolmogorov-Smirnov Z-test. The three size groups have been compared for this purpose: group 1 and group 2, group 1 and group 3, group 2 and group 3. No significant differences could be found.

5.4 Conclusion on manager ratings

Managers rated the importance of companies' attributes - e.g. management focus, financial base, company culture, consequence of decision-making - highest, and the importance of strategic decisions - e.g. acquisitions, alliances & joint ventures, change of business model, and organizational independence - lowest for the relative success or failure of their companies.

From the nine factors that were defined to become part of a base model in chapter 2, four are among the ten most important from the managers' perspectives. The importance of external R&D and resistance to innovation were rated among the ten least important factors. In order to account for the importance of factors that are rated as highly important for the performance of the company, an additional 5 factors are taken into the base model. The additional 5 factors are the ones from the list of the highest-rated factors that have not been covered in research settings such as this one. The additional factors that are included are the focus of top management on technology issues, the company culture, the consequence of decision-making, the extent of decisions, and the distribution of power. Their means of importance are between 4.06 and 4.5 (on a scale from 1 to 5). Factors with a lower perceived importance than 'Distribution of power' with a mean of 4.06 are not included in the base model[11].

Winning and losing does not result in the same managers' conclusions about the priorities of how to successfully manage a company through a technology shock.

Less focus on implementation programs and on the different ways in which innovations are achieved seems to be typically German. The higher focus on finances as compared to other countries in the sample also appears to be typical. German does not mean German-speaking - the Swiss decision-makers have different priorities.

With the exception of resource allocation in the form of allocating employees to the development project of NC/ CNC, company size did not significantly lead to different estimates in what is important for success.

[11] The 'break-off' criterion was chosen to be a mean of 4.06 because there is a clear drop in mean values below this point. 'Distribution of power' is still included because it is perceived as being almost as important as 'cooperation with clients', which is already part of the base model according to chapter 2.

5.5 Managing technology shocks - determinants that distinguish winners from losers

5.5.1 Methods to compare 'winners' and 'losers'

In general, descriptive statistical tools are used in this study to determine how an approximate number of 73 variables helps to distinguish companies that failed in the face of a technological shock from those that succeeded.

The data format in this comparative study is interval-scaled, binary, or metric. When normal distribution can be assumed, t-tests are used to identify significant changes in mean values of interval-scaled and metric data. In the case of binary data and data with insufficient normal distribution, the Kolmogorov-Smirnov Z-test is applied.

The following terminology[12] is used in the overviews: the sample mean $\overline{X}$ is an estimator for μ, the population means; the sample variance s^2 is an estimator for σ^2, the population variance; and the relative frequency k/n an estimator of probability p of the total population. Table A.5.11 of the appendix presents an overview of the t-test statistics and their modifications in homoscedastic and heteroscedastic cases.

Prior to a t-test, every group is tested for an approximate normal distribution using the Kolmogorov-Smirnov test. Therefore, all t-statistics' subgroups are tested for deviations from normal distribution. For $p < 0.05$ a significant deviation from normal distribution is expected and thus the t-test is not applied to these parameters. The use of a t-test would be inefficient and lead to poor approximations. In case of skewed distribution, the Wilcoxon rank-sum test, also known as Mann-Whitney two-sample statistics[13], may be used. Roughly speaking, it allocates ranks to the values of two groups and then compares them. Mann-Whitney statistics work well for a broader range of values containing several rankings. In case of a lesser number of rankings, as in the scale of values from 1 to 5, a different test is favored - the Kolmogorov-Smirnov Z-test. This test relates the maximum difference between the cumulated frequencies in two groups to a z-value. When variables are binary, differences in frequencies of two groups are evaluated by means of the non-parametric Kolmogorov-Smirnov Z-test.

Looking at the distribution curve, it seems that even in cases where the Kolmogorov-Smirnov test rejects normal distribution; the curve is not en-

---

[12] Terminology largely taken from Hildebrand (1986). Equally Toutenburg (2000).
[13] See pp. 427 – 434 in Hildebrand (1986).

tirely different. Moreover, judging such a curve is not easy as most variables range from 1 to 5 in 1-digit increments, i.e. there are only 5 different values. For this reason, the significance of t-test levels of $p < 0.15$ will also be quoted in brackets.

5.5.2 Strength of technology shock and effects of product types

Based on the theory developed about technology shocks in the above sections, different observations depending on the degree of technological change are expected. As for the press group, the technological change was by far less fundamental than for other metal-removing product groups. Thus, deviations in the means observed is expected depending on whether or not the press products group is included. As noted in section 4.5.5, the disruptive effect for press products caused by NC/ CNC was not nearly as radical as for the chip-making machine tools. For this reason, comparisons with and without presses will be made. Tables A.5.12‘a’-A.5.29‘p’ in the appendix show a detailed overview of these results. The set of tables with the letter ‘a’ (i.e. A.5.12‘a’-A.5.14‘a’, A.5.18‘a’-A.5.20‘a’, and A.5.24‘a’-5.26‘a’) contains **all** product groups. The other set of tables with the letter ‘p’ relates to data **excluding** the **presses** group. Based on the criteria of ‘winners’ and ‘losers’, all data series from press manufacturers had previously been classified as ‘winners’. Thus the data in the winners’ column of the ‘p’ tables is expected to change when compared to those including all products. The losers’ columns, however, remain unchanged when compared to tables ‘a’. Although consisting of only two data series, removing the press group data does enhance the differences between ‘winners’ and ‘losers’. This indicates once again that the assumption about the force of technology shocks holds well.

In the following section, the observations are divided into the two groups of ‘winners’ and ‘losers’. The variables describing company attributes, strategic decisions, and operative decisions are examined in three sections. Significant differences between ‘winners’ and ‘losers’ are seen to indicate the importance of special factors. As for similar standard errors of means on the 5-point scale, there are significant variances when the difference in means exceeds 0.5 points.

5.5.3 Significant differences in attributes of 'winners' and 'losers'

Tables A.5.12'a' and A.5.15'p' (see appendix) summarize the differences between companies never having experienced failure and those that have experienced one or more failures during any one of the shock waves 1 to 3. Tables A.5.13'a' and A.5.16'p' express failure and success during shock wave 1 and 2 but not during shock wave 3. Tables A.5.14'a' and A.5.17'p' indicate events that only occurred during shock wave 3 without considering shock waves 1 or 2. Thus, even if a company had failed during one of the other shock waves, it could still be considered a winner. In this sense, the criteria underlying the tables for wave 1 and 2 and wave 3 are more lenient towards failure. The criteria for tables A.5.12'a' and 5.15'p' are stricter as already one failure in wave 1 and 2 would make a company a loser. Significant differences could be observed in for the following variables (an overview is table 5.12 below). Management of companies that never experienced failure during the period of observation emphasized the 'unfiltered access to top management' (CULTFILT). The mean value for this group of winners was 4.55 (max = 5). On the other hand, management of companies that had failed ranged on average between 'impeding changes' and 'actively collecting new ideas' (CULTCOLL). The mean value for losers was 3.43. While winners granted rather unfiltered access of new ideas to top management (mean = 1.95), losers were positioned between 'free access' and 'filtering strongly' (mean = 2.67).

In addition, the priority of technology management in board decisions - or simply a focus on technology management (FOCUSTM) and fast and consequent decision-making (CONSEQUE) seemed to be of substantial importance. Variables clearly defining success thus appeared to be actively collecting new ideas (CULTCOLL), unfiltered access to top management (CULTFILT), focus on technology management (FOCUSTM), and the consequence of decision-making (CONSEQUE). In case the observations about the press group were removed, the involvement of top management in the preparation - or simply focus of involvement (FOCUSINV) (see table 5.12) would also fall into this category.

As explained in the description of the sample, the three observations regarding wave 1 were combined with those of wave 2 for statistical reasons. As the valid number of observations varies for the three shock waves, the means of the total sample changes in tables A.5.12'a' through A.5.15'a'. Unfiltered access to top management (CULTFILT) and the focus on technology management (FOCUSTM) are found to be of significance in waves 1, 2 and 3.

The ability to forecast markets (FORECMAR), resistance to innovation (RESISTAN) and the consequence of decision-making (CONSEQUE) also reach significant levels in wave 1 and 2. Winners attribute significantly more importance to a new market than losers do. Also, their resistance to innovation is significantly lower. These three factors appear to be of no significance in wave 3 (table A.12'a'). However, actively collecting new ideas (CULTCOLL) seems to be of importance in wave 3. The same goes for SALES75 (revenue of companies in 1975) and CULTCOUR (rewards for personal initiative), once the press values are removed.

In the majority of cases where observations by press manufacturers are removed (tables A.5.15'p', A.5.16'p', and A.17'p'), the differences between winners and losers appear much more evident and the levels of significance much higher. Table 5.12 gives an overview of significant differences between winners and losers.

Table 5.12: Overview of significant differences of company attributes

Attribute variables	Waves 1 - 3		Waves 1 and 2		Wave 3	
	'a'	'p'	'a'	'p'	'a'	'p'
FORECMAR		+	*	*		
RESISTAN			*	*		
CULTCOUR						+
CULTCOLL	***	***			***	**
CULTFILT	**	**	+	*	**	*
FOCUSTM	+	*	+	*	+	
FOCUSINV		***				
CONSEQUE	+	*	*	*		
SALES75						+

Significance levels for differences of mean values: *** <= 0.01, ** <= 0.05, * <= 0.1, + <= 0.15.
Waves 1-3 covers the entire observation period from 1955 to 2001.
Waves 1 and 2 relates to events that occurred in wave 1 and 2. Wave 3 accordingly.
'a' (all) reflects results for the entire sample, including the manufacturers of presses.
'p' (without presses) reflects the results for the sample without the 2 manufacturers of presses.

The significant differences are along the lines of the decision-makers' expectations (see table 5.7). The highest ratings went to management focus, expressed by variables FOCUSTM and FOCUSINV, followed by the financial base represented here by SALES75[14]. The third-highest ranking was corporate culture, expressed by variables CULTCOUR, CULTCOLL

[14] In addition, RDABS, RDPERC, RENDABS, and RENDPERC are not significantly different for winners and losers subgroups.

and CULTFILT. In fourth place was fast and consequent decision-making, CONSEQUE, and in fifth place, the ability to forecast both market and technology. The latter, expressed by FORECMAR, however, differed significantly for winners and losers. The only exception was RESISTAN, the resistance to new technology. It was ranked 12th (out of a total of 13), however, it helped make the difference between successful companies and less successful ones in waves 1 and 2. Overall, the top 5 categories (out of a total of 13) that were attributed the highest importance by decision-makers were found to make the difference in determining winners and losers.

5.5.4 Significant differences in strategic decisions of 'winners' and 'losers'

With the exception of modular innovation in the second generation of NC/CNC machines (INNO2MOD) and market entry after dominant design (MARK3RD), the differences are more significant[15] once the press values have been removed (see appendix tables 5.18'a' through 5.23'p').

However, none of the differences are of significance in wave 1 and 2. Yet in wave 3 and waves 1-3, the latter covering the entire time frame from 1955 to 2001, both modular innovation in the first generation of NC machines (INNO1MOD) and market entry after dominant design (MARK3RD) were found to be of significance. Also important to note is that successful companies more often[16] had a similarity to the old model prior to the shock (BMODSIM) and not the same business model as prior to the shock (BMODOLD).

Again, the priorities set by decision-makers in the previous chapter show significant differences. Out of the 8 categories, the following received the highest ratings: type of innovation in the second machine generation, timing of market entry, and type of innovation for the first machine generation. These categories are partially expressed by the variables modular innovation in the first machine generation after the shock (INNO1MOD), modular innovation in the second machine generation (INNO2MOD), architectural innovation in the second machine generation (INNO2ARC), and market entry after a dominant design (MARK3RD). It seems, however, that change or maintenance of business models did not

[15] Since the one-sample Kolmogorov-Smirnov test does not reject normal distribution for OUTOLD in waves 1/2 and 3, the t-test is applied here. For all other interval-scaled and binary variables, the Kolmogorov-Smirnov Z-test is used to analyze differences between winners and losers.

[16] Because of the higher mean in the dichotomous variable BMODSIM.

receive adequate attention. Managers rated this factor in 6th (out of 8), although significant differences do exist for the application of a similar business model after the shock (BMODSIM) and the application of the old business model after the shock (BMODOLD).

Table 5.13: Overview of significant differences of strategic decisions

Strategic decision variables	Waves 1 - 3		Waves 1 and 2		Wave 3	
	'a'	'p'	'a'	'p'	'a'	'p'
INNO1MOD	*	**			*	**
INNO2MOD	*					
INNO2ARC	**	**				
MARK3RD	**	**			*	+
BMODSIM	+	*				
BMODOLD	**	**				

Significance levels for differences of mean values: *** <= 0.01, ** <= 0.05, * <= 0.1, + <= 0.15.
Waves 1-3 covers the entire observation period from 1955 to 2001.
Waves 1 and 2 relates to events that occurred in wave 1 and 2. Wave 3 accordingly.
'a' (all) reflects results for the entire sample, including the manufacturers of presses.
'p' (without presses) reflects the results for the sample without the 2 manufacturers of presses.

5.5.5 Significant differences in operative decisions of 'winners' and 'losers'

As in the comparison of attributes, some of the differences become more significant once the press group was excluded (columns p). Winners were characterized by less growth in the organization that dealt with NC/ CNC (ORGGROW). The reason for this is that winners had committed a larger share of their organization to NC/ CNC to begin with (ORGBEGIN). Losers tried to catch up, but despite over-proportional growth, they never reached the level their successful competitors had regarding NC/ CNC personnel in their organization (ENGTR00). Organizations of winners displayed a span of control of 1:10 in wave 2, almost twice as high as that of the losers (HIERBROA). Although the difference in size of winners' and losers' companies is not significant, the number of employees they hired differed significantly: winners hired less people - on average 9 employees - to acquire critical know-how during wave 2, whereas losers hired around 20 employees (TCHIRE).

The training of business employees for less than two weeks (BSPRC2) came close to a level of significance in wave 2, although only in the case without press manufacturers.

In contrast to the categories 'strategic decisions' and particularly 'company attributes', managers did not prioritize the factors describing differences between winners and losers at a significant level. Resource commitment ranks in 5th place out of 9 options, the importance of new hires ranks 7th and the levels of hierarchy is in 8th place.

Table 5.14: Overview of significant differences of operational decisions

Operational decision variables	Waves 1 - 3		Waves 1 and 2		Wave 3	
	'a'	'p'	'a'	'p'	'a'	'p'
ENGTR00	+	+				
ORGGROW	*	*			+	+
ORGBEGIN			*	*		
HIERBROA			**	**		
BSPRC2				+		
TCHIRE	(+)	(+)	(+)	*		

Significance levels for differences of mean values: *** <= 0.01, ** <= 0.05, * <= 0.1, + <= 0.15.
Waves 1-3 covers the entire observation period from 1955 to 2001.
Waves 1 and 2 relates to events that occurred in wave 1 and 2. Wave 3 accordingly.
'a' (all) reflects results for the entire sample, including the manufacturers of presses.
'p' (without presses) reflects the results for the sample without the 2 manufacturers of presses.

5.5.6 'Early' and 'late' losers

As discussed in section 5.1.6, 6 of the 13 failures during the second shock wave took place at an early stage of the wave, 7 at a late stage. During the third shock wave, 14 of the 25 companies failed at an early stage, whereas 11 failed at a late stage. In this section, we will examine the significant differences between the 'early' and 'late' losers. The parameters are divided into two groups per wave for this purpose - an early group and a late group.

For the second shock wave, significant differences can be seen in table 5.15. No significant differences were found in the strategic variables. Companies that had failed six years after a technological change were more flexible both in their strategic (FLEXBASE) and operative planning processes (FLEXOP) than the ones that had failed during the six years prior to the change. Moreover, the companies that had failed at a later stage rewarded personal initiative higher than their counterparts that had failed earlier on (CULTCOUR). Early losers had fewer employees in the new technology at the end of the observation period (ORGEND). On average, early losers devoted about 50% of their resources to the new technology during the second shock wave; companies that survived longer about 85% of their resources.

Early losers have purchased less R&D from external than late losers (RDPURCR expressed as percentage of revenue). RDPURCHA refers to the same value, simply in absolute numbers.

Table 5.15: Differences in attributes and operational decisions of early and late losers (wave 2)

Variables	Early		Late		Difference of means
	n1	Mean1	n2	Mean2	
Company attributes					
FLEXBASE	6	3.00	7	4.14	1.14***
FLEXOP	5	3.20	7	4.00	0.80+
CULTCOUR	6	3.33	7	4.14	0.81*
Operational decisions					
ORGEND	5	47.60	6	85.00	37.40*
RDPURCHA	5	0.25	6	1.23	2.42+
RDPURCR	5	0.006	6	0.04	0.064+

Significance levels of t-test for differences of mean values:
*** <= 0.01, ** <= 0.05, * <= 0.1, + <= 0.15.

For the third shock wave, significant differences can be seen in table 5.16.

Table 5.16: Differences in attributes and operational decisions of early and late losers (wave 3)

Variables	Early		Late		Difference of means
	n1	Mean1	n2	Mean2	
Company attributes					
FORECTEC	13	3.92	11	2.91	-1.01*
CULTCOUR	12	3.75	10	3.30	-0.45+
RDPERC	5	9.67	8	6.27	-3.40*
ATTRACT	11	3.18	11	4.09	0.91*
Operational decisions					
RDSAVE	7	13.29	7	2.00	-11.29
RDSAVR	7	0.029	6	0.002	-0.027

Significance levels of t-test for differences of mean values:
*** <= 0.01, ** <= 0.05, *< = 0.1, + <= 0.15.

Like in the second wave, there were no significant differences regarding strategic decisions. However, as for the attributes, the picture was quite different during the third shock wave - companies having failed at an early stage were the ones that invested more in R&D (RDPERC), rewarded initiative higher (CULTCOUR), and presented a better forecast regarding technological change (FORECTEC). They were able to save more money by purchasing R&D services externally than companies that had failed at a later stage (RDSAVE and RDSAVR). The early failures were less attractive for new hires (ATTRACT).

During the third shock wave, the trend of becoming involved in new technology was of a different nature, although not to a significant extent. Whereas companies that had survived longer during the second shock wave devoted a higher percentage of their organizations to the new technology, it was now the early losers that devoted 90% to the new technology, the late losers only 77%. This could be an indicator of a unique scenario in the third wave. Early losers in wave 2 seem different in character from those of wave 3. Early losers of shock wave 2 were in an unfortunate position to begin with and failed in the face of technological change. On the contrary, failures during shock wave 3 seemed to be a 'bargain' for other companies to seize the opportunity of acquiring companies with technological potential during a generally dismal economic situation. The sample includes companies interesting as candidates for mergers and takeovers during the earlier part of the third shock wave. The early stage of the second wave, however, shows companies that were just 'doing worse' than other companies that had failed[17].

---

[17] To shed light on this matter, there has to be a way to distinguish between failures caused by bankruptcies and settlements and those caused by takeovers and mergers. One would expect failures due to takeovers to dominate during the early stages of the third wave and those caused by bankruptcies and settlements to dominate during the early stages of the second wave. In order to investigate this assumption further, a scale is constructed according to the rule below. The scale value or index rank is abbreviated INDXRNK and will be used again as endogenous variable in chapter 6 for multivariate analysis.

INDXRNK = number of takeovers and mergers * 0.5 + number of bankruptcies + number of settlements

INDXNK rates bankruptcies and settlements twice as high as takeovers and mergers. INDXRNK7 indicates all failures between 1970 and 1990, INDXRNK9 all failures since 1990. One would expect a higher average of INDXRNK for failures in the second wave compared to the average INDXRNK in wave 3. In addition there is a significant correlation between INDXRNK9

5.6 Summary of the descriptive analysis

Managers rated the importance of factors for company success after a technology shock. Based on their judgment, the group of variables that constitutes a base model for testing in chapter 6 is expanded by management focus, company culture, consequence of decision-making, the extent of decisions, and distribution of power.

There are three observation periods for analyzing firm performance. Wave 1-3 refers to the entire observation period from 1955 to 2001. The observation period for wave 2 is from 1974 to 1989, for wave 3 from 1990 to 2001. The waves are triggered by shock events.

Failure has been defined as bankruptcy, settlement, or takeover. Several companies have undergone a number of failures during the observation period. A rule is applied to identify differences between companies with good performance and companies with a bad performance after the technology shock. A company without any of the three kinds of failures during the observation period is categorized as a 'winner' and a company with one or more failures as a 'loser'. Winners and losers are significantly different from one another in a number of factors that influence success; they are accompanied by significant differences in the mean values of the related success factors. Failures occur both at the beginning of a wave and towards the end. Failures at the beginning of the wave indicate that the theory of product life cycles that predicts failures towards the later stages of a cycle (see section 1.3) requires expansion.

The success factors investigated are from the three blocks of hypotheses: attributes, strategic decisions, and operational decisions. The statements made by the participants on the success factors refer to the time period of the introduction of NC/ CNC, which is for most of the companies between 1970 and 1990, i.e. before the third wave. The results therefore have to be interpreted in the following way: success or failure in wave 2 are put in relation to managerial actions and company attributes during wave 2. This constellation assumes a contemporaneous connection between action and outcome.

Success and failure in wave 3 are also related to managerial actions and company attributes of wave 2. This second constellation shows the long-term effect of actions taken even after the next shock has happened. For a

---

and the early failures during the third shock wave ($p<0.1$). The correlation parameter is 0.34.

complete set of factors that influence the success of the company, this means that results from both wave 2 and wave 3 have to be combined.

Among the attributes, the ones rated most important by managers were also found to be rated significantly differently among winners and losers: ability to forecast (FORECMAR), company culture (CULTCOLL, CULTCOUR, CULTFILT), management focus (FOCUSTM, FOCUSINV), consequence of decision-making (CONSEQUE), and financial base (SALES75). Resistance to innovation (RESISTAN), which was rated as not as important by managers, also showed significant differences.

Among the strategic decision variables, modular innovations in the first and second machine generation after the technology shock (INNO1MOD, INNO2MOD), and architectural innovation in the second machine generation (INNO2ARC) became significant. Characteristic for winners are a rather late market entry (MARKT3RD) and a business model that is different to the old one but still contains elements of the old one (BMODSIM). Losers follow the old business model (BMODOLD). The change of the business model is a factor that was given comparatively little attention by managers.

The operational variables that result in significant differences for winners and losers reflect the following picture. Winners have a higher commitment to the new technology from the beginning (ORGBEGIN). Losers react by increasing resources (ORGGROW) through technical hires (TCHIRE), but they can still never match the amount of attention that winners devote to the new technology (ENGTR00). Winners offer significantly more short training (i.e. 2 weeks maximum) to business employees (BSPRC2). The two highest rated operational decisions, extent of decisions and cooperation with clients, did not show significant differences.

Compared to late losers, early losers of wave 2 are characterized by significantly lower flexibility in their planning processes (FLEXOP, FLEXBASE), have lower rewards for personal initiative (CULTCOUR), had a smaller percentage of resources committed to the new technology and had less R&D outsourced.

Early losers of the third wave, on the other hand, had shown more rewards for initiative (CULTCOUR), higher ability to forecast (FORECTEC), and an optimized R&D (RDPERC, RDSAVE, RDSAVR) in the second wave. They lagged behind the late losers only in attractiveness for new hires.

Thus, the companies that lose early or late are quite different between wave 2 and wave 3. While companies that are in a weaker position to begin with give up early in wave 2, some well-managed companies of wave 2 are the ones that fail in wave 3. A possible explanation for this is that strong companies with the experience of what had happened in the prior shock 2 acquire well-positioned companies to strengthen their own position.

6 Multivariate analysis

6.1 Introduction of scales

In this chapter, multivariate analysis is used to test the three sets of hypotheses introduced in chapters 2 and 3. Ratings by top managers who witnessed the technological change (section 5.4) helped to define a base model for the successful management of technology shocks (figure 6.1).

Figure 6.1: Variable categories in the base model

	Hypotheses block 1: Company Attributes	Hypotheses block 2: Strategic Decisions	Hypotheses block 3: Operational Decisions
Categories in the base model	• **Ability to forecast** • **Resistance to innovation** • **Financial base** • *Distribution of power* • *Consequence of decision making* • Management focus[1] • Company culture[1]	• **Innotype 1** • **Innotype 2** • **Timing of market entry**	• **Resource commitment** • **Cooperation with suppliers and customers** • **Trainings** • **External R&D** • *Extent of decisions*
Candidates for expansion	• Attractivity for new hires • Information flow • Flexibility of plan • In-house manufacturing	• Changing the value chain • Organizational independence • Business model • Alliances & JV • Acquisitions	• Extent of implementation program • New hires • Level of hierarchy

Legend: **Bold** - variable categories with significant results in previous research on technological change (see chapter 2).
Italics - issues suggested by practitioners in the pre-tests (see chapter 3).
[1] - these two items were ranked first and third by managers (see chapter 5).
Source: own depiction.

In chapter 2, three groups of hypotheses were defined from previous research. Each block of hypotheses contains categories which represent one or more variables (see tables A.5.1 through A.5.3). The categories were supplemented through by three items (distribution of power, consequence of decision-making, and extent of decisions) that have been identified in the pre-tests with practitioners in section 3.4. The variable groups management focus and company culture, ranked first and third in importance by managers in chapter 5 were also included in the base model. All other items that were rated better than tenth were already contained in the base model because of previous research or the pre-tests. The other 12 variable groups are candidates for expanding the base model during statistical testing.

Prior to carrying out multivariate analysis, variables are aggregated to scales[1] on the level of variable categories. This follows the assumption that the categories represent hidden constructs. The Spearman correlations are calculated within the variable categories. Cronbach's alpha is used to express the quality of the new constructs. Resulting alpha-values above 0.5 suggest that the constructs are useful aggregates of the underlying variables (Schnell et al., 1995)[2].

6.1.1 Correlations and scales of company attributes

Table A.6.1 in the appendix provides an overview of the Spearman correlations and constructs derived from variables that describe company attributes which have evolved over an extended period of time. A high correlation between the factors indicates that the variable categories can be used to form constructs. The variables can thus be combined into scales, the introduction of which offers several advantages. First, missing values can be imputed with the aid of one or more additional variables of the same scale. Second, scales reduce the influence of errors that stem from flaws in measuring or carrying over values[3].

As seen in table 6.1, FOREC is introduced as a scale consisting of FORECMAR and FORECTEC which denotes the ability of a company to forecast the market and technology. FLEX describes the flexibility of planning on both strategic and operative levels.

Grouping INFCOMP and INFSUP into one variable is an expression of the information flow with external partners (competitors and suppliers). The alpha value for such a scale is 0.55. The question arises whether the

---

1 For the sake of easier handling, the variables are changed to ranges of 1 to 5, as are the resulting constructs. Overall, the introduction of scales reduces the number of variables from 73 to 37.

2 According to Schnell et al, p. 143, values above 0.8 are desirable. Much lower values are accepted in practice.

3 The scales that are found here are similar to those one would construct from a factor analysis.

variable INFINTRN (internal information flow) should be included in this scale because its effects are quite the opposite of INFCOMP and INFSUP and it also has a weaker correlation to the two variables than they do to each other. In order to fully express the construct, INFINTRN is also included in the scale at the expense of a lower alpha value of 0.51.

CULT describes cultural aspects as they are detailed in table A.5.1 of the appendix. It covers the variables CULTCOUR (rewards for personal initiative and courage), CULTCOLL (top management actively collects new ideas from employees), and CULTFILT (free access or filtering of ideas to top management).

FOCUS summarizes management's attitude towards technology management. It includes FOCUSTM, expressing the priority of technology management in board decisions, and FOCUSINV, the involvement of top management in the preparation of decisions about introducing NC/ CNC.

FINANCE represents the financial commitment made to R&D in absolute numbers (RDABS), as a percentage of sales (RDPERC) and with respect to financial base of the company. As such the average absolute and relative profitability measures RENDABS and RENDPERC are included as well as an expression of the company revenue size SALES75 - the company revenue figure for 1975.

Table 6.1: Spearman correlation and scales of attributes

#	Category and variable	In	Construct name	α	Correlation with # within groups			
					2	3	4	5
	Ability to forecast technology							
1	FORECTEC	+	FOREC	0.75	0.57***			
2	FORECMAR	+						
	Flexibility of plan							
1	FLEXBASE	+	FLEX	0.64	0.54***	0.53***	0.29*	
2	FLEXBNUM	+				0.37**	0.54***	
3	FLEXOP	+					0.26+	
4	FLEXOPNU	+						
	Information flow							
1	INFINTRN	+	INF	0.51	- 0.08	- 0.24*		
2	INFCOMP	+				0.30**		
3	INFSUP	+						
	Company culture							
1	CULTCOUR	+	CULT	0.71	0.46***	- 0.36***		
2	CULTCOLL	+				- 0.63***		
3	CULTFILT	+						

Table 6.1 (continuation)

#	Category and variable	In	Construct name	α	Correlation with # within groups 2	3	4	5
	Management focus							
1	FOCUSTM	+	FOCUS	0.67	0.51***			
2	FOCUSINV	+						
	Financial base							
1	RDABS	+	FINANCE	0.72	0.35*	0.80***	0.08	
2	RDPERC	+				0.12	0.45*	0.70***
3	RENDABS	+					0.34	- 0.32*
4	RENDPERC	+						0.78***
5	SALES75	+						- 0.24

Significance levels of Spearman correlation coefficients: *** p<=0.01, **p<=0.05, * p<=0.1, $^{+}$p<=0.15.

6.1.2 Correlations and scales of strategic decisions

Most of the variables for strategic decisions are dichotomous. They are 'dummy variables', a form to be used in the multivariate analysis. For this reason it does not make sense to group them into constructs. The variables OUTOLD (outsourcing of old competence) and INNEW (in-sourcing of new competence) are interval-scaled, although not of a normally distributed shape. The Spearman coefficient of correlation is of no significance here and the alpha value of 0.38 too low as to create a new variable based on a construct.

Table 6.2: Spearman correlation and scales of strategic decisions

#	Category and variable	In	Construct name	α	Correlation with INNEW
	Changing the value chain				
1	OUTOLD	0	-	0.38	0.22 (n.s.)
2	INNEW	0			

Significance levels of Spearman correlation coefficients: *** p<=0.01, **p<=0.05, * p<=0.1, $^{+}$p<=0.15.

6.1.3 Correlations and scales of operative decisions

The seven variables describing the firms' resource commitments are grouped into three scales. First, the commitment of resources in the 1970s forms construct ENG70. It consists of the percentages of employees in technical functions ENGTR70 and employees in business functions

ENGBR70 that had been devoted to the new technology in 1970, i.e. after the introduction of NC and before the CNC technology shock wave. Second, ENG00 is formed following the same logic of ENGTR00 and ENGBR00 - employees devoted to CNC in 2000. Third, ORGPERC is a scale that represents the shares of resources committed to NC/ CNC measured as the percentage of the pre-shock organization and their growth over the years. It contains ORGBEGIN - the percentage of employees dealing with NC/ CNC at the beginning of the shock, ORGEND - the percentage of employees dealing with NC/ CNC at the end of the implementation of NC/ CNC technology, and ORGGROW - the ratio of ORGEND and to ORGBEGIN. ORGGROW represents growth in the area of the organization that deals with NC/ CNC. A detailed description of the resource variables as well as all other variables for operational decisions is given in table A.5.3 in the appendix.

The information about levels of hierarchy (HIERLEV) and the span of control (HIERBROA) are grouped into the scale HIER[4].

CONTROL is introduced as a scale to express to which extent the development of NC/ CNC machines was the result of a stand-alone effort of the machine tool companies or dependent on either customer or supplier input. SUPCONTR and CLICONTR express the variables describing how greatly either clients or suppliers dominated the value chain of the new product. The cooperation with numerical control suppliers was vital for manufacturers of machine tools in order to be able to produce the new NC/ CNC machines. The possibility of creating value decreased as much of the added value was now in the hands of control device manufacturers. Machine tools manufacturers became increasingly dependent on these companies. In this sense, cooperation with suppliers was mostly critical for the success of the NC/ CNC development projects. The variable SUPCOOP takes this fact into account. It is also integrated in the scale CONTROL. A high correlation coefficient ($r = 0.30$) supports this. On the contrary, CLICOOP describes a different quality of cooperation. Although companies of machine tool customers are often substantially larger than those of the manufacturers, customers generally did not get very involved in the de-

[4] For organizations of comparable sizes, it was expected that the number of hierarchy levels HIERLEV and span of control HIERBROA would have a negative correlation. In other words, the broader the span of control in a company, the fewer levels of hierarchy are needed. For the sample under investigation here, though, a positive correlation was found. A possible explanation is that the company size has a dominant influence. The larger the organization, the more levels of hierarchy exist, broadening the span of control. So the resulting scale HIER would be an expression of the size of the company.

velopment of new machine types. They expressed requests and wishes but did not try to influence the development in any specific technical directions. The variable CLICOOP is thus left in its original form. Including CLICOOP would make the alpha value drop to 0.44.

The percentage of employees having gone through formal training programs is summarized in the category TRAINING. It contains variables expressing employees with technical and business training of less than 2 weeks per year (TCPERC2, BSPERC2), between 2 weeks and 6 months (TCPERC4, BSPERC4), and over 6 months (TCPERC6, BSPERC6) as a percentage of all employees.

External R&D spending is expressed by RDEXTERN. It contains the accumulated cost of purchased external R&D services including licenses, patents, and contract research (RDPURCHA) as well as the ratio of RDPURCHA and sales in 1980 (RDPURCR). Also included is the in-house development time, which was saved as an absolute number RDSAVE and as a percentage RDSAVR.

The variables expressing the extent of acquisition of new competence through new hires in technical functions TCHIRE and in business/administrative functions BSHIRE did not show enough correlation to combine them into one variable.

Through the introduction of scales, the number of variables in the group 'operative decisions' was reduced from 27 to 12. The way the variables are grouped is also suggested by a factor analysis of the variables in table 6.3. The highest level of aggregation that can still be interpreted also consists of 10 factors. In a way, this confirms that the assumptions underlying the constructs hold true.

Table 6.3: Spearman correlation and scales of operative decision

#	Category and variable	In	Construct name	α	Correlation with # within groups					
					2	3	4	5	6	7
Resource commitment										
1	ENGTR70	+	ENG70	0.91	0.71***	-0.008	-007	0.32+	-0.06	- 0.41**
2	ENGBR70	+				0.02	0.12	0.20	-0.05	- 0.19
3	ENGTR00	+	ENG00	0.64			0.59***	0.15	0.01	0.30+
4	ENGBR00	+						-0.08	0.006	0.39*
5	ORGBEGIN	+	ORGPERC	0.43					0.46+	- 0.67***
6	ORGEND	+								0.05
7	ORGGROW	+								
Level of hierarchy										
1	HIERLEV	+	HIER	0.54	0.34*					
2	HIERBROA	+								

Table 6.3 (continuation)

#	Category and variable	In	Construct name	α	Correlation with # within groups 2	3	4	5	6	7
Cooperation										
1	SUPCOOP	+	CONTROL	0.54	-0.24*	-0.04	0.30**			
2	SUPCONTR	+				-0.03	0.31**			
3	CLICOOP	0					0.05			
4	CLICONTR	+								
Training										
1	TCPERC2	+	TRAINING	0.70	-0.41**	-0.47	0.64***	-0.29+		
2	TCPERC4	+				0.31*	-0.08	0.78***		
3	TCPERC6	+					-0.01	0.56***		
4	BSPERC2	+						-0.23		
5	BSPERC4	+								
6	BSPERC6	+								
External R&D										
1	RDPURCHA	+	RD-EXTERN	0.71	0.89***	0.89***	0.76***			
2	RDPURCR	+				0.86***	0.83***			
3	RDSAVE	+					0.91***			
4	RDSAVR	+								
New hires										
1	TCHIRE	0	-	0.14	0.16					
2	BSHIRE	0								

Significance levels of Spearman correlation coefficients: *** p<=0.01, **p<=0.05, * p<=0.1, $^{+}$p<=0.15.

6.2 Differences of the scales' means between winners and losers

The variable set after chapter 6.1 now includes new scales that consist of a number of measurables. Differences between the means for the winners and losers groups in the underlying original variables have been analyzed in chapter 5.5. The results are displayed in tables 5.12, 5.13, and 5.14. The mutual correlation of these original variables was used for imputation in order to eliminate several missing values. This section investigates whether the significant differences detected in the original variables would be substantial enough to appear again in the imputed scales. In addition, there could be new significant differences in the scales. These tests are carried out for all data series except for those from the press manufacturers. Cases in which technology shocks were not as severe and the results thus not comparable should not be included.

The predominance of factors from the 'attributes' group and the relative weakness of factors from the strategic decisions group should be tested further. If it is found that a company's attributes predominantly influence its survival of technology shocks, then there is not much that management can do to react to technology shocks on the strategic and operative level. Strategic and operative decisions can be made in the short-term at the time of technological change; attributes, however, are of a nature which cannot be changed quickly. Management focus, culture, consequent decision-making, forecasting ability, etc. have to be learned rather than decided on. They develop within companies over long periods of time and are in many cases related to the personalities of key employees. The survival of companies in the face of technological change would be determined even before the change occurs.

Regarding the strictest criterion distinguishing between 'winners' and 'losers' (i.e. no failure during all three waves 1-3), the scales describing culture (CULT) and management focus (FOCUS) show significant differences like the original variables[5] that they contain. The differences in the ability to forecast did not reach a significant level. As can be seen in table 5.12, they were already at a rather low level of significance with the variable forecasting of markets (FORECMAR) at $p < 0.15$. Unchanged by the introduction of scales were the variables for the consequence of decision-making (CONSEQUE) as well as all of the dichotomous variables covering strategic decisions.

In the variable set resulting from the modifications made in chapter 6.1, these strategic variables still show the same pattern where differences of means are significant as shown respectively in tables 5.13 and A.5.21p in the appendix. These are: INNO1MOD - losers applied a modular type of innovation for the first generation of NC machines significantly more often, INNO2ARC - losers applied an architectural type of innovation for the second generation of NC/ CNC machines significantly more often. Winners tended to have more radical changes in the first machine generation and less fundamental ones in the subsequent machine generations. The other strategic variables with significant differences between winners and losers are: MARK3RD - winners entered the market after a dominant design had become established, BMODSIM - winners had a similar but changed business model significantly more often, and BMODOLD - losers applied the old business model in an unchanged way significantly more often.

[5] The original variables that are contained in the scales are illustrated in table 5.12 of the previous chapter.

Within the group of variables for operational decisions, TCHIRE, which expresses the hiring of technical personnel, remained unchanged by the introduction of scales in chapter 6.1. It thus reappears in the current set of variables with the same differences that are close to a level of significance of $p = 0.15$ (see tables 5.14 and A.5.27p in the appendix). For the scales which were introduced for operational decision variables, only the differences between winners and losers for ENG00 (resource commitment in the year 2000) reached a level of significance. The commitment of employee resources in winner organizations was significantly higher.

For waves 1 and 2, differences in the ability to forecast (FOREC), companies' cultures (CULT), the focus on technology management (FOCUS), the part of the company dealing with the new technology (ORGPERC) and the structure of hierarchy (HIER) are significant. The variables underlying FOREC, CULT, FOCUS, ORGPERC and HIER had already been significant in section 5.5 before scales were introduced (see tables 5.12, 5.13, 5.14). Similar to CONSEQUE and TCHIRE, the variable expressing resistance to innovation (RESISTAN) was not affected by the introduction of scales and still shows significant differences between winners and losers. The scale TRAININGS, which includes the variable BSPRC2 (short-term training of business employees) that has shown significant differences between winners and losers (see table 5.14 and A.5.28p in the appendix), does not show significant results.

In wave 3, only the differences in company culture (CULT) reached a significant level. All others (including FINANCE and ORGPERC containing the individually significant variables SALES75 and ORGGROW as shown in tables 5.12 and 5.14 in the previous chapter) - except the ones not influenced by scales - were insignificant. The variables not influenced by scales were INNO1MOD and MARK3RD, which have the same significant differences between winners and losers as in table 5.13.

Thus the relevant differences between 'winners' and 'losers'- after the original variable set had been modified by the introduction of scales - can be summarized as shown in table 6.4.

The only variable that not only fulfils the strict criterion of no failure (i.e. no takeover, no bankruptcy, and no settlement) for all three waves but also appears significant in waves 1/ 2 and 3 is CULT, which describes the company's culture. The other differences are wave-specific. This means that different characteristics play a role over time in further analyses. Wave 3 as the subsequent technology shock to waves 1 and 2 is still influenced by the previous managerial decisions. The variable INNO1MOD illustrates this. Although INNO1MOD describes the modular type of innovation in the first generation of NC machines which emerged around the first or second technology shock wave, INNO1MOD still had a significant influence on how companies performed in the third shock wave. The past decision on the degree of innovation in reaction to the previous, second wave has a differentiating effect for winners and losers in the third shock wave. Companies that had decided to create their first generation of NC machines by making modular changes to their existing machines were more often among the losers in the subsequent technology shock.

Table 6.4: Significant differences between winners and losers after the introduction of scales

Variable	Waves 1 - 3	Waves 1 and 2	Wave 3
FOREC^S		**	
RESISTAN		*	
CULT^S	***	**	***
FOCUS^S	+	***	
CONSEQUE	*	*	
INNO1MOD	**		**
INNO2ARC	**		
MARK3RD	**		+
BMODSIM	*		
BMODOLD	**		
ENG00^S	+		
ORGPERC^S		*	
HIER^S		**	
TCHIRE	(+)	*	

Significance levels of differences of mean values: *** p<=0.01, **p<=0.05, * p<=0.1, $^+$p<=0.15.
^{S} this variable is a scale.

Table 6.4 illustrates some of the central results of the analysis of differences between the 'winners' and 'losers' groups of companies. It is expected that some[6] of the factors will also show significant results in the multivariate analysis of section 6.6.

6 Probably not all of the factors will appear significant in the multivariate analysis. Correlations as shown in tables A.6.1 through A.6.3 in the appendix make the results of the multivariate analysis deviate from the univariate analysis.

6.3 'Early' and 'late' losers

The differences for early and late failures in wave 2 and wave 3 are now investigated in order to find out how the introduction of scales and imputation affected the results as compared to the original variables in section 5.5.6. The scales for wave 2 containing variables that showed significant differences in section 5.5.6 yield significant differences once again. That means that the difference between winners and losers has not been affected greatly through aggregation. The only exception is company culture (CULT). In its aggregated form, it does not show any significant difference between winners and losers although its elements prior to aggregation had been significantly different for winners and losers.

The results for wave 3 are rather surprising. None of the scales that contain variables whose differences have been found significant individually (see table 5.16: FORECTEC - ability to forecast technology, CULTCOUR - encouragement of initiative, RDPERC - investment in R&D, RDSAVE and RDSAVR - savings by externally purchased R&D) appears again with significant differences[7]. Instead, the ones that do appear are training programs (TRAINING) and the variable that describes the situation of a manufacturer being controlled by a supplier (CONTROL). The limited sample size of companies under investigation, the heterogeneous distribution of data entries and missing values contribute to the effect that there are significant differences for scales whose underlying original variables have not shown significant differences in section 5.5.6.

An overview of the significant data series following the introduction of scales is provided in table 6.5.

Table 6.5: Overview of significant differences between early and late losers

Variable	Wave 2	Wave 3
FLEX^S	+	
ATTRACT		*
ORGPERC^S	+	
CONTROL^S		*
TRAINING^S		+
RDEXTERN^S	+	

Significance levels of differences of mean values: *** p<=0.01, **p<=0.05, * p<=0.1, $^+$p<=0.15.
^{S} this variable is a scale

After having introduced the scales, there was no similarity to be detected between the 'early' and the 'late' losers in both waves. This is another indicator of why the waves should be treated individually in the further analysis.

6.4 Determining causal relations through multivariate analysis

The correlation between the resulting set of variables is displayed in tables A.6.1 - A.6.3 in the appendix. This set of variables is the one that will be used for multivariate analysis following the introduction of scales. The variables' contribution to different measures of success is being investigated in the subsequent sections.

The following information is used to operationalize the endogeneous success and failure of companies:

1. The scale INDXRNK, i.e. the average weighted numbers of failures - bankruptcies and settlements to avert bankruptcy - are rated with the weight '1.0' and mergers and takeovers with the value '0.5'. The scale was already introduced in chapter 5.5.6, where it was used to investigate differences between companies that failed either early or late.
2. The duration of survival under risk. Taking survival time into consideration when investigating organizational mortality has become a frequently-used instrument[8] in literature on population ecology and industrial organization.

An additional possibility of measuring success was used in chapter 5: a dichotomous variable which groups companies by attributing 'winners' the value '1' and 'losers' the value '0'. Using a logit regression approach would be one alternative of dealing with dichotomous dependent variables (Fienberg, 1977; Kennedy, 1996). Using the normal cumulative function is the other alternative - the probit model. Given the small sample size at hand, it is not feasible to test which alternative is more appropriate. The dichotomous variable was the focus of the descriptive analysis. The dichotomous variable determining 'winners' and 'losers' appeared in three ways: first, the strict criterion in which companies with only one failure since 1955 were already categorized as 'losers' and the second being the 'core' shock wave 2 from 1974 to 1989, where companies were termed 'losers' if

[7] The variable ATTRACT does show the same results as in table 5.16 because it was not affected by the introduction of scales.

[8] See Barnett (1990), Klepper (1999).

they experienced a failure during this period of time. Those that did not experience any failure were automatically in the 'winners' category irrespective of whether they failed before or after the said time frame. And third, the 'current' shock wave 3 as of 1990 where the terminology follows a logic similar to wave 2.

Market share information was collected as another measure of success. Market share is an observable with an ex-post perspective[9]. Market shares decline with the revenues of companies that did not manage the technology shock successfully. In fact, there is a significant correlation of market share changes[10] between 1980 and 1990 and the dichotomous variable for the second wave. The correlation between the change in market share between 1970 and 1980 and winners and losers is not significant. As wave 2 has been defined to last from 1974 until 1989, one could indeed draw the conclusion that market share changes are delayed a few years following a technology shock.

Different multivariate approaches will be used to test for causal relations between company success and the set of exogenous variables. The multivariate analysis, which aims at finding the relevant factors for a successful management of technology shocks, splits the analysis into two parts.

1. Multi-nominal linear regression is used on the failure scale INDXRNK described above covering the entire observation period between 1955 and 2001. By virtue of its definition, the scale offers more differentiation in the sense that 'losing several times is worse than losing once' than the simple distinction between 'winners' and 'losers'.
2. The Cox approach uses survival time data in order to test the factors that influence the probability of survival during waves 2 and 3. In this case, the Cox approach is preferred over a logit model since it contains additional time series information. The Cox model is also used to identify changes between waves 2 and 3.

[9] Decision-makers from industry made similar observations: Frank Baumbusch (Ex-Pittler): 'Wenn Verluste gemacht werden, ist es meistens schon zu spät' ('Once losses are made, it is usually already too late'). Thus the results of unsuccessful management during a disruptive period are felt after a delay (sometimes up to the present day).

[10] True for both market shares within home markets and the global market.

6.5 Displaying survival

Kaplan-Meier survival rate statistics (Toutenburg, 2000) are applied to calculate a cumulative probability of survival. The starting point contains all companies. Companies that fail are registered with their year of failure[11] and the probability curve declines. There were no failures up to the year 1971 in the sample. The first relatively substantial drop in survival probability could be observed around 1975. That year was the beginning of shock wave 2 (see the descriptive analysis). The second major decline began in the 27th year, which is in 1982, around the time when the late effects of wave 2 hit the industry. The most drastic fall was in 1992 at the beginning of shock wave 3. The graph below (figure 6.2) provides an overview of how failure frequencies in the sample developed between 1955 (year 0) and 2001 (year 46).

Figure 6.2: Survival probability determined by first-time failures

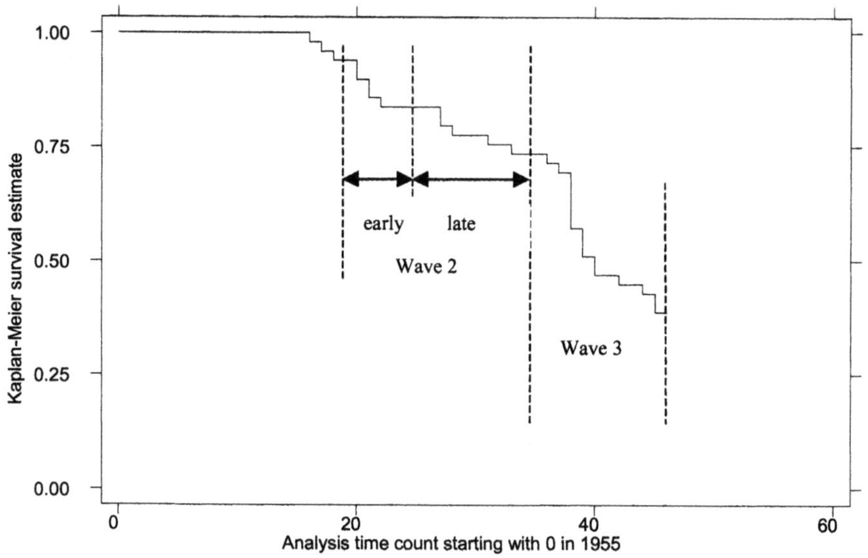

Failures attributed to the second wave are only those that occurred between 1974 and 1989. The period of analysis for the second wave starts in 1974. Failures after 1989 are already attributed to the third wave.

[11] In this type of statistics, only one failure per company - i.e. the first failure - can be taken into account. The 'Zombie' phenomenon as described in chapter 5.1.4 is not taken into account here.

Out of the three companies that failed during the first wave, only one company failed again both in wave 2 (4 years after the shock trigger) and in wave 3 (5 years after the shock trigger). With only one difference in failure, the curve shape of wave 2 would correspond to the curve above between years 19 and 34. But as the survival rate chart for wave 2 does not yield new information if the failure cases of wave 1 are suppressed, it is omitted here.

Figure 6.3 gives detailed information about wave 3 when all of the companies - even the ones that had already had failures before 1990 - are included in the start sample again. The clock for all of the companies in the sample is reset[12], i.e. even if a company had failed during the previous wave it would be included again in the starting panel of companies at time 0 in the year 1990.

Figure 6.3: Survival probability determined by failures in the 3rd wave

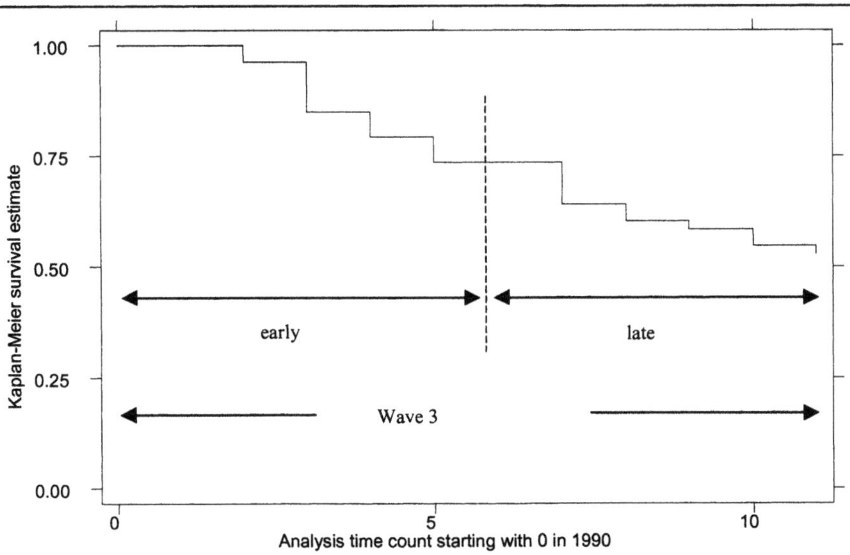

The effect described in section 5.1.4 as the 'Zombie' phenomenon, that companies can have multiple failures, leads to a slightly different curve shape for wave 3 in figures 6.2 and 6.3. The reason is that seven companies experienced multiple failures during both waves 2 and 3. The survival rate curve in figure 6.3 shows a slightly clearer pattern than in the previous figure - there are no changes in survival rate between years 5 and 7. The 3rd

[12] For an example of dealing with time variables and moments, see Amburgey, Kelly and Barnett (1993).

wave is now also split into an 'early' and a 'late' wave like wave 2 in figure 6.2.

6.6 Analyzing the determinants of survival

In the following section, the hypotheses obtained from literature in chapter 2 and statements by experts during the pre-tests and the interview series detailed in chapters 3, 4, and 5 are tested. In the process, 9 variables were taken from previous research and 6 from expert interviews with managers, which were incorporated into a basic model for statistical testing. Due to the relatively large number of variables, the procedure applied analyzed the variables in groups: first, the variables describing company attributes were tested, then the strategic decision variables, and finally the operational decision variables. In turn, the following sections are structured into 'attributes', 'strategic decisions', and 'operational decisions'.

As described in section 6.4, a regression on INDXRNK is used to identify factors determining management success for the entire period from 1955 to 2001 punctuated by the three shocks. The weighted number of failures seems appropriate to account for the different levels of success companies experienced during this period.

To obtain a detailed analysis of waves 2 and 3[13], the Cox model (Cox et al., 1984) is used to detect causal relationships between the probability of survival and the covariates. In this case, the Cox model appears to be superior. In addition to incorporating information on success and failure it also makes use of the duration of success following the shock trigger and preceding the next failure. This approach thus builds on more extensive data than others[14] would.

For each hypothesis group, the period 1955 to 2001 is analyzed by means of a regression model. In addition, waves 2 and 3 are investigated using the Cox model to uncover wave-specific influences on company success. The process is started with the hypotheses group on company attributes and is repeated for the strategic decision hypotheses and the hypotheses on operational decisions.

---

[13] There are not enough failures in the sample for wave 1.

[14] Logit is based on information on success or failure, ordered logit on failure categories. Regression was not applied to the wave-specific analysis because the number of failures that can be attributed to waves 2 or 3 yields less information than combined information on the duration of survival and success or failure as in the case of the Cox model.

The exogenous variables used to model INXRNK and survival time are not time-dependent. They are assumed to have the same value in all time periods.

The tests begin with the base model that builds on previous statistical work and the variables prioritized by managers. The base model is tested; non-significant variables are dropped from the model and supplemented by additional factors one by one. The additional factor which results in the largest increase in χ^2 of the model[15] is included in the model. The factor that yields the second best result of χ^2 increment is also tested with the new model. Thus the optimized model consists of those factors from the base model that are at a level of significance and of one or two additional variables that were candidates for expansion.

Finally, the findings from the three groups of hypotheses are combined into models on how to successfully surpass technology shocks: one describing the entire time span from 1955 to 2001, one describing only wave 2, and one only wave 3. For the interpretation of results, it is important to know that performance after shock 3 is influenced by the decisions that were made in the face of shock 2. The values of the variables used for modeling wave 3 are the same as for the other models. The timing of these variables precedes shock wave 3. The results therefore have to be seen in one large context of shock waves following each other. With the knowledge that certain factors have an effect on company performance even after a subsequent shock wave, an anticipatory and diligently-managed company would have paid attention not only to factors that are specific to wave 2 but also those that will have a positive or negative impact in the next shock wave.

6.6.1 Multivariate analysis of the variables that represent company attributes

The variable FINANCE (financial base of the company and financial commitment to R&D) suffers from missing data. Even though it is a scale of 5 variables, it consists of only 35 observations. The scale FINANCE is constructed from resource figures, among them expenses for R&D and the sales figures of 1975; thus a high correlation between the number of employees and total revenues can be expected. Indeed there is a high correlation ($p < 0.0001$) between total number of employees, total revenues, and financial figures. The data are thus used to impute the variable FINANCE for all 58 entries. The imputed variable is named IMPFINANCE.

[15] All being at a level of significance.

IMPATTRACT is the imputed form of ATTRACT (attractiveness of the company for new hires) with the aid of BSHIRE (number of new business hires) and TOT00 (the number of total employees in the company in the year 2000). FOCUS (management focus on technology) and CONSEQUE (consequence of decision-making) show significant correlations with 5 and 6 of the other 10 scales respectively and remaining original variables that describe company attributes (see table A.6.1 of the appendix). They are imputed with the help of the entire variable group of attributes. These attributes can be detected already before the shock. It is assumed that they are not time dependent.

The basic regression model for INDXRNK consists of seven variables - three which had shown significant results in previous research on technological change, two that had been tested before but in a different context, and two variables that were included because of prioritization by managers. The financial base (IMPFINANCE), resistance to innovation (RESISTAN), and the ability to forecast markets and technologies (FOREC) are the variables that were derived from past studies in the field of technological change. Consequence of decision-making (IMPCONSEQUE), cultural aspects (CULT), the distribution of power (POWER) and the focus by top management (IMPFOCUS), are the ones that were identified as the most critical factors by top management and are included in the base model test. The model is estimated as follows:

$$\text{INDXRNK} = f(\beta_0, \beta_1\,\text{FOREC}, \beta_2\,\text{RESISTAN}, \beta_3\,\text{CULT}, \beta_4\,\text{IMPFOCUS}, \beta_5\,\text{IMPCONSEQUE}, \beta_6\,\text{IMPFINANCE}, \beta_7\,\text{POWER}),$$

where β_0 is the constant factor and β_1 through β_7 are coefficients of the variables.

Only those coefficients are retained that indicate a level of significance less than $p = 0.15$. The resulting reduced base model contains two variables: management focus (IMPFOCUS) and consequence of decision-making (IMPCONSEQUE). The factors FOCUS and CONSEQUE had already shown significant results for the comparisons of the means for winners and losers in section 6.2 (see table 6.4)[16]. The variable CULT, however, no longer shows any significant results in the regression.

Subsequently, all other variables are added one by one while their increment to R-squared is registered. As none of the variables show significant contributions, the reduced base model consisting of two variables remains unchanged and thus represents a 'best fit' version. The significance of variables when they are added one by one to the base model and their increment to R-squared is shown in the appendix A.6.5.

---

[16] FOCUS and CONSEQUE presented significant differences for 'winners' and 'losers' in the descriptive statistics of chapters 5.2 and 6.2 (table 6.4).

The results have to be interpreted in this way: The number of failures (as expressed in INDXRNK) is lower when technology management is a high priority on the managerial agenda and when decisions are made quickly and implemented consistently.

Table 6.6: Effects of company attributes on failure (1955 - 2001)

Variable	Coefficients of reduced base model
Management focus (IMPFOCUS)	-.47** (.20)
Consequence of decision-making (IMPCONSEQUE)	-.20+ (.13)
R-squared	.198
RMSE	1.00
Number of observations	58
F	6.77
P of model	<0.002

Significance levels of differences of mean values: *** p<=0.01, **p<=0.05, * p<=0.1, +p<=0.15.
Note: OLS Regression model, standard errors in brackets

The results might change if the single shock waves were analyzed separately, i.e. shocks 2 and 3. The Cox model was applied because it makes use of additional information - survival time following the triggering shock.

In the Cox regression model, the hazard is assumed to be

$h(t) = h_0(t) \exp(\beta_1 x_1 + \ldots + \beta_k x_k)$.

The time dependent $h_0(t)$ is called the baseline hazard[17]. The base model that is put to test consists of the same 7 variables as in the regression model. The variables that describe company attributes and are included in the model are set to be time-invariant. In fact, it is assumed that they describe the companies already before the shock occurs. The values of the variables do not change depending on whether they are used in a model for the wave 2 or wave 3 periods[18].

The model that results after the first test for wave 2 consists of only one variable and a constant. The consequence of decision-making (CONSEQUE) has proven to be a factor differentiating 'winners' and 'los-

---

[17] The Cox model calculates estimates of $\beta_1, \ldots, \beta_k$ which in turn are used to estimate the cumulative hazard $H_0(t)$ and the survival function $S_0(t)$. The procedure is similar to one of regression. A base model is being estimated first and increments of other variables are tested step by step. See Toutenburg (2000): The relation between H_0 and S_0 is expressed in terms of $S_0(t)=\exp(-H_0(t))$.

[18] Figure 5.2 in section 5.1.1 describes the timeframe of the different variable groups. The start point at which the three variable groups can be measured differs (attributes first, operational decisions last) but the values for each variable are assumed to stay the same over time.

ers' in the descriptive analysis of sections 5.4 and 6.2 (see table 6.4). IMPCONSEQUE represents a hazard ratio[19] of 0.63** with a standard error of 0.15[20]. The result of the Cox model has to be interpreted like this: Hazard ratios smaller than 1 indicate a positive effect on the duration of survival, ratios larger than 1 a negative effect. The other attribute variables are tested consecutively for their relevance (see appendix A.6.6).

The model consisting of consequence of decision-making (IMPCONSEQUE) and attractiveness of the company in the eyes of potential new hires (IMPATTRACT) has a χ^2 of 7.38, which is lower than the χ^2 of 7.63 of the model consisting of consequence of decision-making (IMPCONSEQUE) and ability to forecast (FOREC). Therefore the most suitable model (best fit) for two variables consists of consequence of decision-making (IMPCONSEQUE) and ability to forecast (FOREC). The combination of consequence of decision-making (IMPCONSEQUE), ability to forecast (FOREC), and attractiveness of the company for new hires (IMPATTRACT) does not result in coefficients that are at a level of significance.

Table 6.7: Effects of company attributes on hazard ratios in wave 2 (1974 - 1989)

Variable	Hazard ratios of reduced base model	Hazard ratios of best fit model
Consequence of decision-making (IMPCONSEQUE)	.63** (.15)	.67* (.16)
Ability to forecast (FOREC)		.63** (.14)
CHI-squared	3.68	7.63
Number of observations	58	54
P of model	< 0.1	< 0.05

Significance levels of differences of mean values: *** p<=0.01, **p<=0.05, * p<=0.1, $^{+}$p<=0.15.
Note: MLE Cox model, standard errors in brackets

19 Which is equivalent to the estimator of exp(β), i.e. the hazard ratio for a 'one-unit change' in x.

20 The corresponding estimator of the coefficient is -0.45** (with a standard error of 0.23).

The same procedure[21] is applied in wave 3 using the Cox model (see appendix A.6.7). The reduced base model after the first test for significance consists of the imputed scales for financial strength and finances committed to R&D (IMPFINANCE) and the distribution of power (POWER). After the candidates for expansion have been tested on the model, a best fit model[22] emerges that consists of three variables. Additional results include the focus of management (IMPFOCUS). Including the variable for company culture (CULT) which reached a level of significance in a three variable model with IMPFINANCE and POWER does not result in a significant model of four coefficients. The model with three significant coefficients and the highest χ^2 contains IMPFINANCE, POWER, and IMPFOCUS.

Table 6.8: Effects of company attributes on hazard ratios in wave 3 (1990 - 2001)

Variable	Hazard ratios of reduced base model	Hazard ratios of best fit model
Financial base (IMPFINANCE)	2.71** (1.12)	2.31** (.89)
Distribution of power (POWER)	1.59+ (0.48)	1.61+ (.51)
Management focus (IMPFOCUS)		.60* (.17)
CHI-squared	6.08	9.33
Number of observations	46	46
P of model	< 0.05	< 0.05

Significance levels of differences of mean values: *** p<=0.01, **p<=0.05, * p<=0.1, $^{+}$p<=0.15.
Note: MLE Cox model, standard errors in brackets

6.6.2 Multivariate analysis of the variables that represent strategic decisions

The categorical information about possible strategic actions is used in this section to test the survival probability displayed in the Kaplan-Meier graphs (see charts in figure 6.4 below). INNO1 expresses the degree of in-

[21] It has to be noted that the exogenous variables are time independent. Their measured values relate to the time period of shock wave 2 (for strategic and operational decisions) or preceding shock wave 2 (for company attributes). The variable values are assumed to remain the same even during wave 3. Their operationalization is described in tables A.5.1 through A.5.3 in the appendix. Imputation and introduction of scales in previous sections of chapter 6 does not change the fact that the variables are not time dependent.

[22] The factor with the best level of significance and highest increments in chi-squared is added to the model. The second best factor is tested on the new model and included if the hazard ratios are all at a level of significance and if chi-squared increases.

novation represented by the first generation of NC machines as compared to previous non-NC machines. INNO2 reflects the degree of change represented by the subsequent NC/ CNC machine tool generation. MARKT differentiates between three times of market entries (first to market, before a dominant design emerged, and after a dominant design emerged) for NC/ CNC machines. These three variables are related to the second technology shock wave. Thus for analysis purposes survival data can be used for the entire observation period (1955 - 2001) or for wave 2 only (1974 - 1989). The three variables relate explicitly to waves 1 and 2 so that no 'artificial' interpretation for wave 3 is made.

The INNO1 and INNO2 data represent four degrees of innovation - incremental, modular, architectural, and radical. The top two graphs below show that careful innovation (i.e. incremental) during the first generation of products based on a new technology was a decision clearly associated with a higher likelihood of company survival. Incremental in this context means that companies kept adding functionalities to the machines, introduced the actual NC rather late (mainly between 1975 and 1980) and the CNC only a few years later. For 50% of the companies introducing CNC in a subsequent generation of machines (expressed by INNO2), this represented modular innovation, for the other 50% radical innovation. These findings reflect the situation of 12 of the companies in the sample: six of them experienced incremental innovation, the other six modular innovation during the first generation of machines. The findings are supported by what the fifth and the sixth graphs below suggest - late market entry leads to higher chances of survival than early entry[23]. In this sense the findings are contradictory to some of the initial assumptions about market entry (Klepper et al., 1990; Shapiro et al., 1999; Smith et al., 1988) and support the findings of Reidenbach et al. (1986), which propose that a longer and more intensive preparation effort increases the probability of success. The window of opportunity, which supposedly occurs in fast-changing industries before a dominant design establishes itself in the market (Christensen et al., 1998), would be around 1980, a time when Japanese CNC machines became overwhelmingly successful. The window of opportunity hypothesis holds good for the 6 companies that innovated incrementally until they entered the market with their CNC machines around 1980 - in this case, it also applies for a whole industry that follows longer cycles of change than what was assumed by Christensen.

In figure 6.4, six charts displaying the effect of different strategic decisions regarding the type of innovation and market entry are shown in two

[23] For a similar finding see section 5.3.4: 'Winners'entered the market late significantly more often than 'losers'.

variants. One variant is displayed on the left-hand side; it shows the entire observation period from 1955 until 2001. The other one shows the observation period after shock 2 in 1974 until the third shock in 1990. It is displayed on the right-hand side.

In order to be able to make statements about the timing and degree of technological change, combinations of INNO1 and INNO2 have to be investigated. The logic and examination of the data suggests that INNO1 and INNO2 are not independent of each other.

Four fundamental combinations[24] are feasible:

- A company adapts only slightly (i.e. incrementally or modular) to the new technology in the first machine generation following the technology shock (INNO1). In the second generation of machines following the shock (INNO2), the company undertakes a major technological advance (i.e. architectural or radical change), bringing it up to a higher technological level (this combination is labeled Inno low high in figure 6.5)
- A company undertakes the major technological advance in the machine generation directly following the shock. The second generation of machines represents only slight changes compared to the previous machine generation (Inno high low).
- A company never really implements the new technology. Its degrees of innovation are comparatively small in the first and second machine generations (Inno low low).
- A company undertakes a major technological advance in the first machine generation but changes the concept of the machines entirely in the second generation (Inno high high).

[24] The two most frequent combinations are - first, low degree of innovation during the first generation and high degree during the subsequent generation, and second, high degree of innovation for INNO1 and low degree of innovation for INNO2.

Figure 6.4: Survival probability differentiated by type of strategic decisions

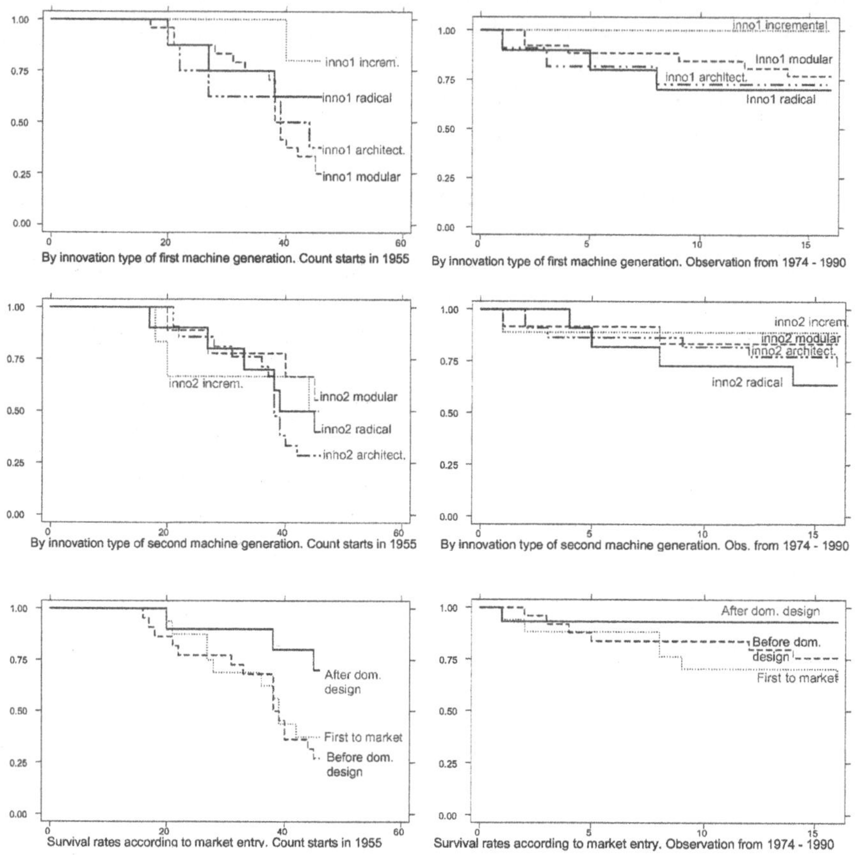

Kaplan-Meier graphs are used to identify potentially different impacts of the combination of innovation types on firm performance. The diagrams below reflect the different combinations of the types of innovation. Intersecting lines show that - with the exception of the 'low - low' combination that appeared for the period after the second shock and before the third shock in six companies - there is no preferred combination.

The hypothesis that targeting a new market with architectural innovation would contribute to success (Christensen et al., 1998) cannot be supported here. In this case, this hypothesis would translate into architectural change in the first generation (INNO1). This variant of innovation type does not demonstrate better company performance.

Figure 6.5: Survival probabilities differentiated by combinations of innovation type

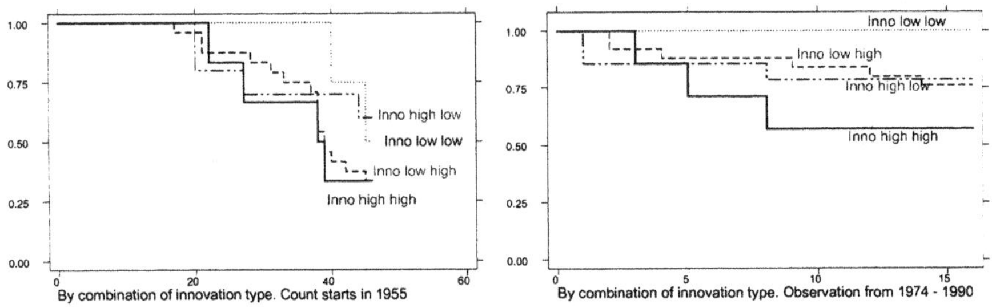

The appendix of this chapter includes the analysis of 3 more variables - the degree of organizational independence and separation (ORG), the type of business model (BMOD), and the variable (ALLIANCE), which indicates whether an alliance was initiated in the context of acquiring NC/ CNC know-how. The Kaplan-Meier graphs are displayed in table A.6.4 in the appendix. The following observations can be made:

The graphs for ORG reflect the fact that only four companies had organized their NC/ CNC activities in separate units. Two of these attempts did not survive independently. Equally, only four companies had established their NC/ CNC in a business unit within the organization. None of the hypotheses on organizational independence obtained from literature in section 2.3.2 (table 2.2) can be tested using this type of data.

There are 3 graphs in table A.6.4 that display the influence of the business model (BMOD) on company performance. Five companies using new business models had the most successful performance at the time of a technology shock. 15 companies with modified business models experienced a higher likelihood of success than those that did not change their business models at all. They used a strategy for the new market that contained elements from their previous strategy. Thus these findings seem to support Cooper et al. (1992) and Link (1987).

The intersecting lines for ALLIANCE indicate that alliances have no influence on the successful management of technology shocks in this case.

Multivariate models are set up based on this knowledge - i.e. in what manner the type of innovation, market entry timing, organizational independence, change of business model as well as the existence of alliances influence the probability of company survival following a technology shock. In the multivariate models the ordinal information contained in INNO1, INNO2, MARKT, ORG, BMOD, and ALLIANCE is expressed in the form of dummies. For the variables in the base equation - derived from the statements in chapter 2 - the optimal combination of dummies to express categorical information is found by trying out the different permutations[25]. The base model contains INNO1INC, INNO1MOD, INNO1ARC, INNO2INC, INNO2MOD, INNO2ARC, MARKT1ST, and MARKT3RD[26]. In this set of variables, the three strategy factors rated highest by top managers are included: the type of innovation of the first and second machine generations and the timing of market entry (see figure 5.5). After testing for statistical significance, the base model is reduced to two variables, INNO2MOD and MARKT3RD. This base model is described in table 6.9. The other strategic variables are being tested for their suitability in the model (appendix A.6.8).

The information whether another company has been acquired and integrated (TAKEOVER), representing only 40 observations, is imputed with the help of timing of market entry (MARKT) and the degree of outsourcing of old capabilities (OUTOLD). The correlation of these variables could originate from the fact that companies entering markets late resort to acquiring resources and vice versa: in some cases acquisitions can make companies enter new markets. The second correlation is based on a tendency to outsource old competence in the course of acquiring a new one.

The most suitable model (best fit) includes modular innovation in the second machine generation after the shock (INNO2MOD), comparatively late market entry after the establishment of a dominant design (MARKT3RD), and the unchanged old business model (BMODOLD). Late market entry and an old business model (BMODOLD) were already seen to yield significant results for the analysis of differences between 'winners' and 'losers' in section 6.2: Old business models were characteristic for the 'loser' group whereas late market entry was characteristic for 'winners'. Negative coefficients indicate a reduction of the number of failures. Positive coefficients indicate a growing number of failures.

[25] 2 dummies are sufficient to express information contained in a scale of 3 elements; similarly 3 dummies are enough for a scale of 4 elements.

[26] The variables stand for modular innovation, architectural innovation, and radical innovation for the first NC machine tool generation, modular innovation and architectural innovation for the second generation of NC/ CNC machines and first to market entry and market entry after a dominant design.

Table 6.9: Effects of strategic decisions on failure (1955 - 2001)

Variable	Coefficients of reduced base model	Coefficients of best fit model
Modular innovation for the second machine generation (INNO2MOD)	-.64** (.30)	-.60* (.31)
Market entry after dominant design (MARKT3RD)	-.91** (.32)	-.60+ (.36)
Old business model (BMODOLD)		.68** (.32)
R-squared	.225	.264
Adj. R squared	.194	.214
RMSE	1.02	1.01
Number of observations	53	48
P of model	< 0.01	< 0.01

Significance levels of differences of mean values: *** p<=0.01, **p<=0.05, * p<=0.1, $^{+}$p<=0.15.
Note: OLS Regression model, standard errors in brackets

Similar to the approach taken for the company attributes, the period of shock waves 1 and 2 are investigated using Cox models. Like in the OLS regression model (for example in table 6.9), the basic Cox model consists of variables describing type of innovation in the first and second machine generations and timing of market entry. Testing this basic Cox model does not result in any coefficients that are statistically significant. As a result, all parameters including the ones designated for model expansion (see hypotheses block 2 in figure 6.1) are tested independently. Variables that are tested additionally describe changes in the value chain, organizational independence, the business model, acquisitions, and alliances and joint ventures. Three variables show significant results: late market entry (MARKT3RD), similar but changed business model (BMODSIM), and the old, unchanged business model (BMODOLD). The parameter that best fits is the one that expresses the old business model (BMODOLD).

Table 6.10: Effects of strategic decisions on hazard ratios in wave 2 (1974 - 1989)

Variable	Hazard ratios of best fit model
Old business model (BMODOLD)	4.04* (3.11)
CHI-squared	4.39
Number of observations	52
P of model	< 0.05

Significance levels of differences of mean values: *** p<=0.01, **p<=0.05, * p<=0.1, $^{+}$p<=0.15.
Note: MLE Cox model, standard error in brackets

The result of the Cox model has to be interpreted like this: Hazard ratios smaller than 1 indicate a positive effect on the duration of survival, ratios larger than 1 a negative effect. It follows that the influence of the old business model on the probability of survival is clearly adverse: A company employing an old business model after a technology shock is likely to fail sooner.

The following constellation is of high relevance for the successful management of the third technology shock ('wave 3'). The base model is reduced to three variables: modular innovation in the second machine generation (INNO2MOD), the timing of market entry - in this case first to market (MARKT1ST), and the late market entry (MARKT3RD). Both market entry variables have a positive effect on survival probability.

Table 6.11: Effects of strategic decisions on hazard ratios in wave 3 (1990 - 2001)

Variable	Hazard ratios of reduced base model		Hazard ratios of best fit model	
Modular innovation for the second machine generation (INNO2MOD)	.26**	(.15)	.25**	(.15)
First to market (MARKT1ST)	.52+	(.24)	.34**	(.18)
Market entry after dominant design (MARKT3RD)	.14***	(.10)	.10**	(.10)
Old unchanged business model (BMODOLD)			3.3**	(1.2)
CHI-squared	15.87		20.25	
Number of observations	48		43	
P of model	< 0.01		< 0.001	

Significance levels of differences of mean values: *** p<=0.01, **p<=0.05, * p<=0.1, +p<=0.15.
Note: MLE Cox model, standard error in brackets

From the other covariants, two business model variables 'similar but changed business model' (BMODSIM) and 'old business model' (BMODOLD) show significant results. The increase in χ^2 is greater for BMODOLD (see appendix table A.6.10). As a result, BMODOLD is included in the model that fits best after these two steps.

6.6.3 Multivariate analysis of the variables that represent operational decisions

The process is repeated for operational decisions. The percentage of resources of the organization dealing with NC/ CNC (ORGPERC) is imputed with the aid of the total number of employees in 1990 (MITTOT90) and the total number of employees in 2000 (MITTOT00). Variables of operative decisions suffer most from missing data. An additional scale is introduced to reduce the number of variables that could potentially appear in the regression equation while at the same time increasing the number of data points. It thus reflects the correlation between organizational variables and inspires theoretical interpretation. The scale BIGORG basically represents the commitment to employ resources on NC/ CNC. Low values of BIGORG express low commitment, high values high commitment. It consists of the engineering resources of the company in 1970 (ENG70), in 2000 (ENG00), of the percentage of the organization dealing with NC/ CNC (ORGPERC), the hierarchical structure (HIER- many levels of hierarchy and a large span of control), the hiring of technical (TCHIRE) and business resources (BSHIRE). The resulting alpha value is 0.55. A high value of BIGORG indicates large organizations with a high commitment to resources and the willingness to hire resources externally.

The high correlation between amount of research done externally (RDEXTERN) and the resources saved through outsourcing (RDSAVE), the saved resources as a percentage of total company resources (RDSAVR), the purchase of R&D results and services (RDPURCH), and the same value in percent (RDPURCR) is used for the sake of imputation.

With a regression approach on the variables expressing hypotheses block 3 - operational decisions - (see figure 6.1), only the extent of decision (DECISION) comes close to a level of significance of $p = 0.15$.

No significant relationships could be detected when the Cox model was used for waves 2 and 3. In the final model where attributes, strategic decisions and operational decisions are combined, only the extent of decisions (DECISION) from the operational decisions group will be included in the test.

Along with the variables that have been detected within the two other groups, section 6.7 describes the test for joint significance.

6.7 A combined model of managing technology shocks

The findings concerning company attributes, strategic and operational decisions expressed in the 'best fit' models in the paragraphs above are now used to construct a model that covers all aspects identified in sections 6.6.1 through 6.6.3.

First an OLS regression model that describes the probability of failure for the entire observation period from 1955 to 2001 is being constructed. It makes a statement about the variables that influence company success and failure during a period where three shock waves hit the industry.

Second, two more models cover the second and the third shocks individually. These two models are Cox models that cover the periods 1974 - 1990 and 1990 - 2001 respectively. By analyzing two consecutive shocks individually, it becomes clear which factors are relevant in the preceding shock wave and which ones in the subsequent shock wave.

The exogenous variables used to model INXRNK and survival time are not time-dependent. They are assumed to have the same value in all time periods.

The time-independent exogenous variables are used to describe three endogenous variables that relate to different time periods. In the OLS regression model that describes the entire time period, the weighted number of failures - INDXRNK - counts the failures between 1955 and 2001. In the second model, the Cox equation is applied to explain the duration of survival after the second shock between 1974 and 1989. In the third model, the Cox equation refers to the survival times of companies after the third shock wave between 1990 and 2001.

In section 6.6 the variable candidates for a regression that aims at modeling the period 1955 to 2001 have been identified. They are IMPFOCUS, IMPCONSEQUE, INNO2MOD, BMODOLD, DECISION, and MARKT3RD. The latter (MARKT3RD) is eliminated as it does not reach a notable level of significance.

Table 6.12: Overall model - effects on failure (1955 - 2001)

Variable	Coefficients of best fit model	
Management focus (IMPFOCUS)	-.58***	(.22)
Consequence of decision-making (IMPCONSEQUE)	-.26*	(.14)
Modular innovation for the second machine generation (INNO2MOD)	-.55*	(.30)
Old business model (BMODOLD)	.48+	(.30)
Extent of decisions (DECISION)	.41**	(.17)
F	6.12	
R-squared	.43	
Adj. R-squared	.36	
RMSE	.92	
Number of observations	46	
P of model	< 0.001	

Significance levels of differences of mean values: *** p<=0.01, **p<=0.05, * p<=0.1, $^{+}$p<=0.15.
Note: OLS Regression model, standard error in brackets

In the OLS regression, variables with negative coefficients express a reduction of the endogenous variable INDXRNK; variables with positive coefficients express an increase. Thus the interpretation for this 'best fit' model in table 6.12 is as follows: the number of failures is significantly lower for companies that emphasized technology management (IMPFOCUS), experienced modular product innovation during the second generation of machines (INNO2MOD), did not follow the old business model (BMODOLD), and whose decision scope was rather incremental

(DECISION) but decisions were made quickly and implemented consistently (IMPCONSEQUE). While focus on technology management and consequence of decision-making are attributes of companies which are usually determined already before a shock occurs, the decision on type of innovation, business model and extent of decisions are determined by management after a technology shock.

All these factors - with the exception of BMOD - were identified by managers as being important, managers had rated them among the top 7 most important factors (see section 5.2, table 5.6). The importance of changing an old business model was not appreciated by managers. In addition to descriptive statistics (see table 6.4), multivariate OLS regression analysis has shown again that managers underestimated the importance of new or modified business models. Cooper and Smith (1992) and Link (1987) had found in case studies and a quantitative study on innovation that the business model for the new market should be close to the previous business model. The findings here put this into a different perspective: the business model has to be different from the old one in order to reduce the probability of failure.

Past research on the focus of management has suggested that respecting technology sufficiently in board discussions and involving top management in technology questions has a positive effect on the success of companies (see table 2.1). This phenomenon also holds true for the technology shocks in the machine tool industry.

The findings of Christensen et al. (1998) that targeting new market segments with an architectural innovation tends to be more successful than those that innovate in components cannot be confirmed here. In contrast, it was demonstrated that modular innovation leads to success.

Other factors which had shown significant results in previous research on technological change[27] did not appear to be statistically significant here. Consequence of decision-making (IMPCONSEQUE) and the extent of decisions (DECISION) are factors which have not been analyzed in this context in the past. They were suggested by practitioners and found to be significant. This result suggests that these two issues should be respected more strongly in future research and recommendations for practitioners.

The consequence of decision-making (IMPCONSEQUE), ability to forecast (FOREC), and an unchanged business model (BMODOLD) are the candidates for the final model of wave 2. This combination results from statistical testing in the previous sections. The model contains factors that

[27] Ability to forecast, resistance to innovation, financial base, timing of market entry, resource commitment, cooperation with suppliers and customers, trainings, and external R&D.

significantly influence the duration of survival after the second shock wave until the third shock wave. Table 6.13 below describes the outcome of taking these three factors into a Cox model with the duration of survival after shock wave 2 as the exogenous variable.

Table 6.13: Overall model - effects on hazard ratios in wave 2 (1974 - 1989)

Variable	Hazard ratios of best fit model	
Consequence of decision-making (IMPCONSEQUE)	.68+	(.17)
Ability to forecast (FOREC)	.52**	(.14)
Old business model (BMODOLD)	4.71*	(3.82)
CHI-squared	13.75	
Number of observations	51	
P of model	< 0.01	

Significance levels of differences of mean values: *** p<=0.01, **p<=0.05, * p<=0.1, +p<=0.15.
Note: MLE Cox model, standard error in brackets

The Cox model is interpreted as explained in section 6.6.1: Hazard ratios smaller than 1 indicate a longer duration of survival. Hazard ratios larger than 1 result in a shorter survival time.

In the second wave, the ability to forecast technology and market increased the survival time of a company. Price (1996), Cohen and Levinthal (1990), Cooper and Smith (1992), Cooper (1979), and Cooper and Kleinschmidt (1986) discussed the positive effects of the ability to forecast in their works (see section 2.3.1). Now this effect has been identified in a quantitative test in the context of a technology shock.

The two variables IMPCONSEQUE and BMODOLD, which are also contained in the model for the entire time period 1955 - 2001 (see table 6.12), have a significant influence on the survival duration of the second shock wave. The consequence of decision-making increases the duration of survival statistically whereas applying an old business model after the technology shock reduces the survival time.

The potential candidates for the Cox regression in the third wave are the attributes management focus (IMPFOCUS), financial basis (IMPFINANCE), distribution of power (POWER) and the strategic decision variables modular innovation in the second machine generation (INNO2MOD), first market entry (MARKT1ST), market entry after dominant design (MARKT3RD), and the old unchanged business model (BMODOLD). Both timing of market entry expressions do not reach a level of significance. The other variables constitute the best fit model (table 6.14).

Table 6.14: Overall model - effects on hazard ratios in wave 3 (1990 - 2001)

Variable	Hazard ratios of best fit model
Management focus (IMPFOCUS)	.55** (.16)
Distribution of power (POWER)	1.89** (.61)
Financial base (IMPFINANCE)	1.91* (.71)
Modular innovation for the second machine generation (INNO2MOD)	.28** (.18)
Old business model (BMODOLD)	2.54* (1.28)
CHI-squared	20.7
Number of observations	42
P of model	< 0.001

Significance levels of differences of mean values: *** p<=0.01, **p<=0.05, * p<=0.1, +p<=0.15.
Note: MLE Cox model, standard error in brackets

Besides BMODOLD, which also appears in the model for shock wave 2, the model for wave 3 contains two more variables - IMPFOCUS and INNO2MOD - that appear in the overall model for the entire period 1955 - 2001 (see table 6.12). In addition, the model for wave 3 is made up of a variable for the distribution of power. As the hazard ratio is larger than 1 for this variable, a higher value of POWER means a statistically shorter survival time. According to the operationalization of this variable in the appendix of chapter 5 (see table A.5.1), a lower value of POWER means that the ideas of lower levels in the hierarchy get implemented. This variable was suggested by practitioners in the pre-tests. Thus research has not placed enough attention on the effective distribution of power in a company so far.

An unexpected result is the hazard ratio for IMPFINANCE of 1.91. In the researched sample, financial strength does not necessarily contribute to the survival time of the companies under investigation.

6.8 Summary of the multivariate results

Figure 6.6 below provides an overview of the factors that constitute models of successful management of technology shocks. The effect of the respective factor on success or failure is indicated by symbols, i.e. '+' is used to show a significant positive effect on the company, '-' to show a significant negative effect. The factors needed to lead a company successfully through the three[28] shocks of technological change are listed in the column called 'All waves'. These factors are: focus by top managers on issues of technology management (IMPFOCUS), consequence of decision-making

[28] Actually only two waves 2 and 3, due to data limitation.

(IMPCONSEQUE), type of innovation of the second generation of machines with a new technology (in this case modular innovation - INNO2MOD), application of the old business model to the new technology (BMODOLD), and the extent of decisions (DECISION). While the first three factors have a positive effect on the performance of companies, an old business model (BMODOLD) and too radical decisions (DECISION) reduce the chance of success following a technology shock.

In the overall model, variables describing the timing of market entry drop below the level of significance. As for analysis dealing solely with the hypotheses of strategic decisions, late market entry had a positive influence on company performance; a finding that contradicts the results of much previous research (Christensen, 1997; Christensen et al., 1998; Klepper et al., 1990; Shapiro et al., 1999; Smith et al., 1988; Tushman et al., 1986). Late market entry signifies in a sense longer and more careful preparation (Reidenbach et al., 1986).

The consequence of decision-making and the scope of decisions have been suggested by practitioners during the pre-tests to this research and not by literature. Past work on technological change did not reflect these aspects sufficiently. This research suggests that more weight has to be attributed to these two factors in practical management and future academic work.

All except one of the factors in the overall model had been rated among the most important by managers in chapter 5. The issue which was underestimated was the importance of changing the business model after a technology shock. Several authors had dealt with the issue of the business model already, although with less articulated findings. Whereas Link (1987) and Cooper and Smith (1992) had stressed similarities of old and new business models after innovations, the finding here is more imperative. The new business model has to be different from the old one in order to increase the probability of success.

While the positive effect of managers focusing on technology issues was in line with what was expected from past research findings, finding that modular innovation following a technology shock increases the probability of success was certainly unexpected. Christensen et al. (1998) had suggested architectural innovation. On the contrary, it was found that modular innovation in shock wave 3 leads to success.

The research setup enabled a detailed analysis of two subsequent shock events. Two models were shaped to express the influence of different variables on the survival probability after shock wave 2 and shock wave 3. The underlying Cox logic is built on survival duration as endogenous variable in contrast to number of failures in the overall OLS regression model 'All waves' for the overall observation period 1955 - 2001. Still, the results of the models for shock waves 2 and 3 turn out to be complementary and contain the factors that are found in the overall 'All waves' model.

So for wave 2, two factors that are contained in the overall model have proved significant. These are the consequence of decision-making (IMPCONSEQUE) and the adverse influence of an old business model (BMODOLD). In addition the company's ability to forecast (FOREC) has been identified to be a significant element of the model for survival during shock wave 2.

In the subsequent technology shock - referred to as shock wave 3 - attributes and decisions that influenced the company during the second wave continue to have an influence on company performance. The factors that were found to have a significant effect are the managers' focus on technology management (IMPFOCUS), modular innovation in the second machine generation (INNO2MOD) and again, the adverse influence of an old business model (BMODOLD). These three variables are also contained in the overall model 'all waves'. BMODOLD turned out to be of relevance in all situations. This factor appears to be significant in both detailed models for waves 2 and 3 as well as in the overall model for the entire time period under observation.

In addition, the financial situation of the company (IMPFINANCE) and the distribution of power within the company's hierarchy (POWER) are contained in the model for wave 3. In contrast to expectations (Smith et al., 1988; Tilton, 1971), a strong financial situation in the second wave leads to an increased failure rate in wave 3.

If the implementation of the new technology in wave 2 is based on lower levels of the hierarchy rather than on higher levels of the organization, the probability of success following a further technology shock is higher. The variable 'distribution of power within the company hierarchy' (POWER) was proposed by practitioners during the pre-tests of this research. Its importance was underestimated in research up to this point. It seems that when lower hierarchical levels in a company drive technological change, the organization survives technological change better.

So all three factors suggested during the pre-tests - consequence of decision-making, extent of decisions, and distribution of power - turned out to be essential elements in modeling the management of technology shocks. These three variables have been introduced in management of technological change for the first time. The results suggest that they should be applied in future research as well.

In the overall model for wave 3, variables describing the timing of market entry drop below the level of significance. In the analysis dealing solely with the hypotheses of strategic decisions, late market entry had a positive influence on company performance as well as first to market. No indication has been found for a 'window of opportunity' at a time shortly before a dominant standard has become established. As mentioned before, this is not in line with previous findings in other industries (see Christensen, Suárez, and Utterback 1998), but it might be typical for industries with longer development cycles such as the machine tool industry.

Resistance to innovation, resource commitment, cooperation with suppliers and customers, training, and external R&D, all candidates for success factors in the base model, turned out to have no significant influence.

So the multivariate analysis results in findings that challenge previous research. From ten variables that made up the base model, which were motivated by previous research on radical technological change, only one - the ability to forecast - appeared in the way past research implies. Three of them - late market entry, modular type of innovation, and financial base[29] - had an effect opposite to that found by previous researchers. The others did not reach a level of significance.

Three factors were added to the model for the first time in the context of technology shocks based on practitioner recommendation. All three - consequence of decision-making, the extent of decisions, and the distribution of power within the company hierarchy - turned out to be highly relevant in explaining the successful management of technological change.

The focus of management which had been suggested by literature but was not yet statistically tested for its positive effect on company success could be added to the list of relevant factors.

Finally, the necessity of changing the business model clearly established itself as an important condition following a technology shock. This fact has to be respected more strongly by researchers and practitioners.

The practical interpretation could read as follows. Quick decision-making and a consistent implementation process were just as crucial to surviving shock wave 2 as the ability of the company to forecast market activity and technology correctly. In addition, the business model had to be changed.

Management decisions and a company's situation not only affect the current constellation but also severely influence the company success in future radical technological changes. The timing for the variables precedes technology shock 3, so the following interpretation applies. Wave 3 as the subsequent shock wave is influenced by company attributes and management decisions during the previous shock wave 2. If management was focused sufficiently on technology during wave 2, if the old business model had been changed, if the operative management was influenced strongly by

[29] In contrast to previous research, late market entry, modular innovation, and a weak financial base increase the probability of success.

lower levels in the company hierarchy and if the second machine generation was a modular innovation, then the success of the company in future technological changes is likely to be greater. In this case, great financial strength can even be viewed as adverse.

Figure 6.6: Resulting determinants for the management of technology shocks

	All waves (1955 - 2001)	Wave 2 (1974 - 1989)	Wave 3 (1990 - 2001)
Attributes	• Management focus (+)		• Management focus (+)
	• Consequence of decision-making[1] (+)	• Consequence of decision-making[1] (+)	
		• **Ability to forecast (+)**	
			• ***Financial base (-)***
			• Distribution of power[1] (-)
Strategic decisions	• ***Modular Innotype 2 (+)***		• ***Modular Innotype 2 (+)***
			• **(First to market) (+)**
	• ***(Late market entry) (+)***		• ***(Late market entry) (+)***
	• *Old business model (-)*	• *Old business model (-)*	• *Old business model (-)*
Operational decisions	• Extent of decisions[1] (-)		

Legend: **Bold** - variable categories with significant results in previous research on technological change.
Italics - results are in contrast to previous thinking in literature.
[1] - these three issues had been suggested by practitioners and had not been included in previous research.
(+) significant positive effect on company performance, (-) significant negative effect on company performance
Source: own depiction.

If management is looking for a list of priority issues with respect to technology management, it should make sure that these ten aspects have been addressed properly. As shown in figure 6.6, there are five aspects - so-called company attributes - that require a longer period to establish, four strategic decisions, and one operational decision that are to be made when a technology shock strikes.

7 Conclusion and outlook

A framework consisting of two dimensions can be used to identify technology shocks: one dimension being the effect of technological change on a company's competence, the second dimension being product architecture, which usually translates itself into performance parameters. Applying the framework to the introduction of numerical controls within the machine tool industry reveals three shocks between 1955 and 2001, which changed the industrial structure and caused established firms to fail[1]. If the first appearance of a commercial product on the market defines the beginning of a technology shock, then the first shock took place in 1955, the second in 1974 and the third in 1990. Researching these three technology shocks results in implications for research and management of which only the most important ones should be named here.

It can be expected that technological change in the future will create technology shocks again and trigger the same processes that were already observed three times in the history of the machine tool's industry.

Technology shocks trigger product life cycles, which are characterized by management's reactions to the change. In addition to what was suggested by theory, an increase in failures not only occurred at the end of a shock wave, but also at the beginning. This indicates that the theory of product life cycle (PLC) and technological change requires an extension and an additional phase might have to be introduced into the PLC phase scheme. An over-proportional number of failures have not only been detected during the last phase of 'cutthroat competition', but also directly following the occurrence of a new technology. Managerial actions in 'strong' companies (e.g. takeovers to strengthen the company's own position) - in anticipation of a PLC pattern - could explain the increase in failures of 'weak' companies.

Both the long time span between the shocks and the rather slow adoption of the numerical control technology by the industry contribute to the fact that the chosen industry is termed 'slow-moving' as opposed to 'fast-moving' industries. Factors that have an impact on the performance in fast-moving industries - which had been the focus of quantitative research re-

---

1 The framework was used to identify future potential shocks.

garding technological change so far - are not congruent with factors that influence success in slow-moving industries. While, for example, early market entry (Smith *et al.*, 1988) or entry during the 'window of opportunity' (Christensen, 1997; Christensen *et al.*, 1998) is beneficial in fast-moving industries, firms entering the market after a dominant design had become established performed better in slow-moving industries.

This analysis investigated factors that influenced the success of companies in three groups - company attributes that have developed over a longer period of time, strategic decisions, and operational decisions. Two company attributes appear to be dominant. One is top management's dedication of priority to and the extent of involvement in technology management. The other is the consequence of decision-making and implementation - a factor suggested by managers for investigation.

Management can influence the success of a company after a shock by means of strategic decisions. The decisions regarding the degree of innovation in machine architecture and the change of a business model accompanying the new products have a significant influence on a company's performance after a shock. The machine generation that follows the first generation of machines with a new technology ought to be the result of modular innovation rather than of incremental, architectural, or radical innovation. Continuing with an unchanged business model despite new technology makes failure more likely. In addition, the scope of decision-making has a significant impact on companies' success: aiming at incremental changes rather than far-reaching changes resulted in better performance.

A focus on technology management, consistent and timely decision-making and implementation and the change of the business model alongside a new technology contribute positively to the success of companies.

Management focus and the change of the business model had not been tested econometrically in connection with technological change prior to this study[2]. The results of this research confirm hypotheses from conceptual work and studies in different environments as expected.

The consequence of decision-making as well as the extent of decisions are novel[3] variables explaining company performance after a technology shock.

Additional factors become relevant depending on the observation period. For example, during the 'second' shock wave (1974 - 1989), a company's ability to develop sensitivity for new markets and new products was

---

2 Proposed conceptually and tested in a more general context of innovation.

3 Practitioners proposed these variables during pre-tests.

of significant relevance. The ability to forecast new products and markets had already been subject to tests prior to this study.

Success during the subsequent 'third' technology shock wave (1990 - 2001) was still influenced by managerial actions that were undertaken already before in response to the second shock. In addition to what was found in the overall model, it turned out to be significant that the implementation of the new technology during the second wave had generally followed the initiative of lower levels of hierarchy within the company. This factor is also new in the discussion of technological change[4]. Contrary to what was expected based on previous research, the financial strength of a company during the second shock wave reduces the chance of success after the occurrence of the third shock.

An important implication for management is that one factor should receive higher attention on the managerial agenda. While managers acknowledge the importance of other factors[5] seen in the analysis as having a significant influence on success, they tend to underestimate the relevance of what was found here to be the most dominant factor for success - the change of the business model.

The results obtained in this study are drawn from a limited number of observations in one industry and are restricted by the inherent complexity of the machine tool industry. In order to extend the external validity of the statements, the theoretical possibility of transferring the approach to other industries could be used. In the course of this research, studies were made regarding manufacturers of the telecommunications infrastructure[6] and an experiment was carried out within a biotech company. These studies are not to be discussed here, but do indicate that the methodology and the factors that potentially influence performance can also be transferred to other industries that evolve comparably slowly. The fact that three variables out of ten in the final model are entirely novel in this context and the coefficients of four of the variables contradict findings from previous research indicates two things. First, the area of radical technological change is still rather unexplored. Second, it appears useful to treat research of radical technological change in industries with long development cycles as a sepa-

4 The implementation of this variable within the research variable is the result of discussions with managers.

5 Flexibility of plan appeared rather insignificant as compared to its importance during the second wave.

6 The focus of these studies was the transition from analog to digital switching systems. The industry currently faces a technology shock due to substitution of circuit-switched technology by packet-switched technology and the convergence of data, voice, and video.

rate field. The results are likely to be quite different from the findings made in fast-moving industries that have been in the focus of interest so far.

The limitations of internal validity caused by the difficulty of obtaining a larger-sized sample could be avoided by repeating similar investigations in industries where a higher number of observations can be obtained. Besides enabling higher levels of significance, it might allow for the discovery of more subtle effects.

Appendix of chapter 1

A.1.1 Rebecca Henderson's framework for defining innovation

Radical and incremental innovations serve as the extreme points along two dimensions. The first dimension is the degree of change in a product's core concept. The second is the way that components of a product are linked together. A component is defined as a portion of a product which 'embodies a core design concept'. Thus the product is either regarded as a system or a set of components. Due to differentiation and possible variations of parameters, two further types of innovations result: modular and architectural innovations.

Figure A.1.1: Henderson's 'framework for defining innovation' and Afuah's 'hypercube'

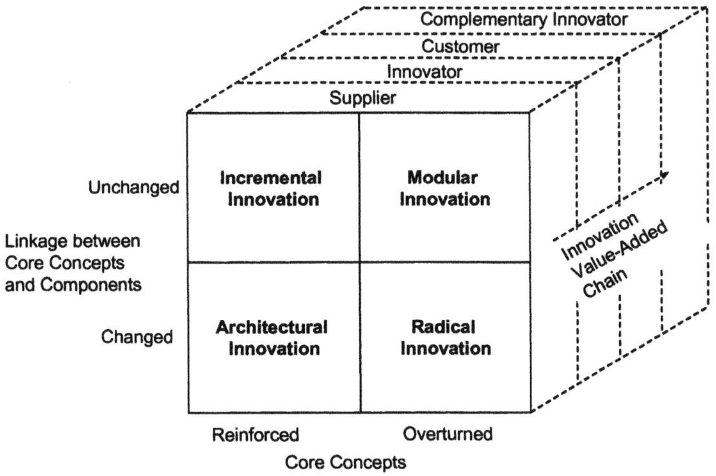

——— Henderson, Rebecca M., and Kim B. Clark. 1990. 'Architectural Innovation: The Reconfiguration of Existing Product Technologies and the Failure of Established Firms.' Administrative Science Quarterly 35: p.12.

------- Afuah, Allan N., and Nik Bahram. 1995. 'The Hypercube of Innovation.' Research Policy 24: p.53.

Afuah and Bahram (1995) added a third dimension to the matrix: by adding the value chain as an axis, they emphasize the fact that technological change can have varying effects on companies depending on their respective roles in the value chain: what might be modular for the innovator could be radical for the supplier. All kinds of combinations are theoretically possible.

A.1.2 Clayton Christensen's trajectories of the impact of sustaining and disruptive technological change

Sustaining technologies improve the most important performance parameter on a steady trajectory. The technical improvements necessary to achieve this may be incremental but also radical in a technological sense. Disruptive technologies result in lower performance on the traditional scale. However, through sustaining innovations it can eventually reach a level at which it can fulfill market requirements from low to high end. Underlying is the assumption that technologies develop faster than market requirements, so that the performance of the established product is eventually too high for the market. Characteristics of disruptive products are that they are cheaper, simpler, and offer lower margins.

Figure A.1.2: Clayton Christensen: the impact of sustaining and disruptive technological change

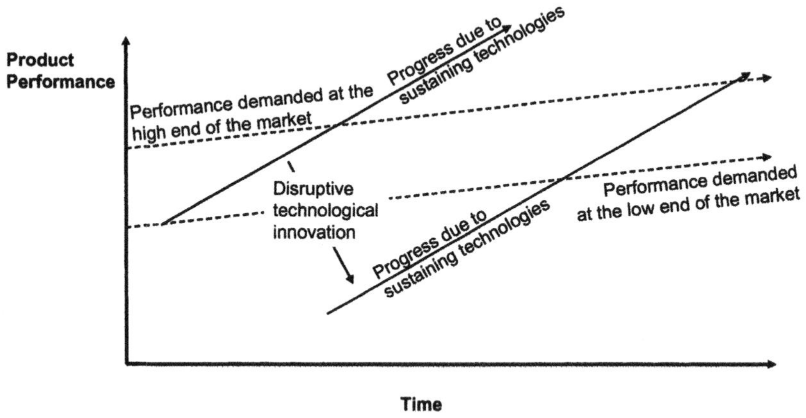

Source: Christensen, C. M. (1997). 'The innovator's dilemma: when new technologies cause great firms to fail.' Boston, Harvard Business School Press: p.XVI.

A.1.3 Abernathy and Clark's 'transilience map'

In order to express the role of innovation in competition, major competitive elements form two groups: the first containing technology/ production related factors, the second containing market/customer related factors. Among the technology/ production related factors are the design of technology, production system, skills, supplier relations, equipment, and experience base. The market/ customer related factors include relationship with the customer base, improved or new customer applications, distribution and service channels, demand of customer knowledge, and modes of communication with customers. The two groups of factors are axes of the two-dimensional matrix displayed below.

Figure A.1.3: Abernathy/ Clark: 'transilience map' to identify the role of innovation in competition

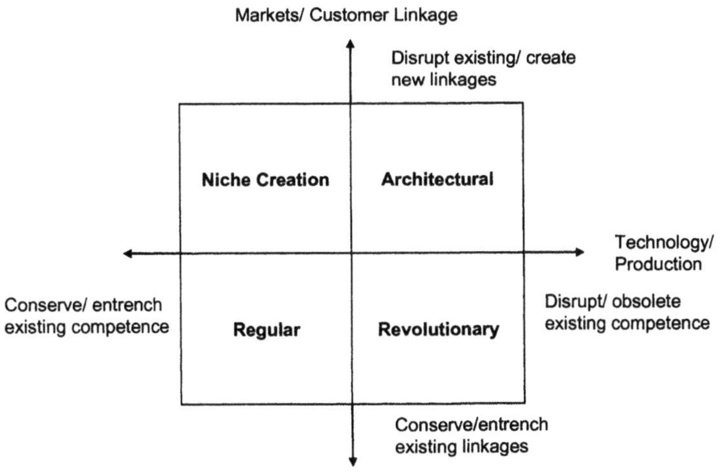

Source: Abernathy, W. and K. Clark (1985). 'Innovation: Mapping the Winds of Creative Destruction.' Research Policy 14: p.8.

Appendix of chapter 3

Figure A.3.1: Screenshot - entry page with links to questionnaires

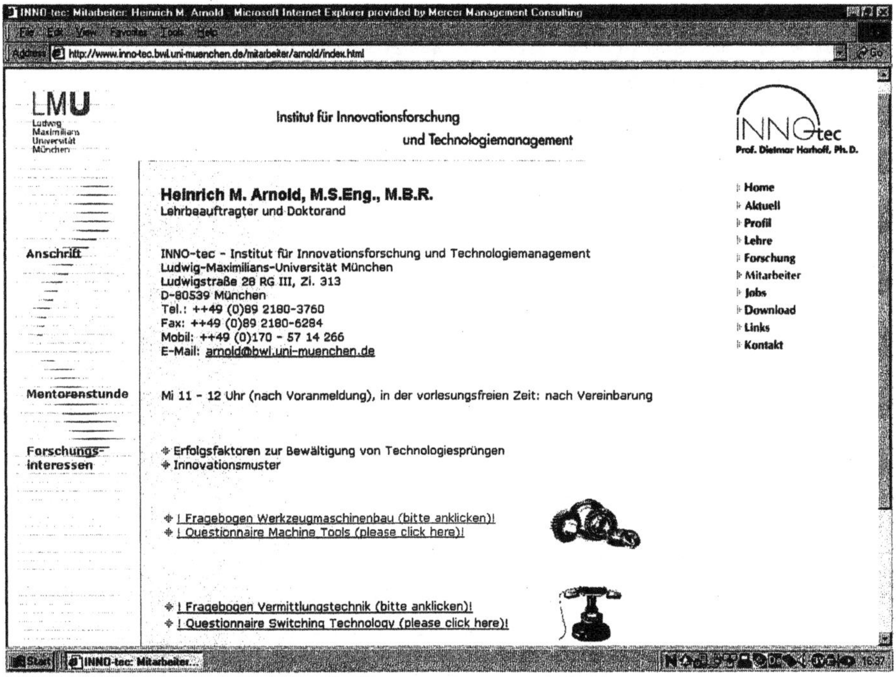

Above is the page for the machine tool and telecommunication infrastructure industries with links to the English and German questionnaires. The links are indicated on the left of the telephone and gear symbols. Below are the two-page headers of the English version. Examples of how participants were asked to enter the information (in the white data fields) are shown in screenshots A.3.1 and A.3.2.

Once the participants had pressed the 'submit' button at the end of the questionnaire, the data was automatically returned in the form of an e-mail and the workings of the transfer module tested; i.e. whether the data had been entered in the appropriate data fields. The latter were given self-explanatory names for easy location of data after transfer to other software packages (in our case Microsoft Excel®, STATA®, and SPSS®).

Figure A.3.2: Screenshot - English questionnaire machine tools

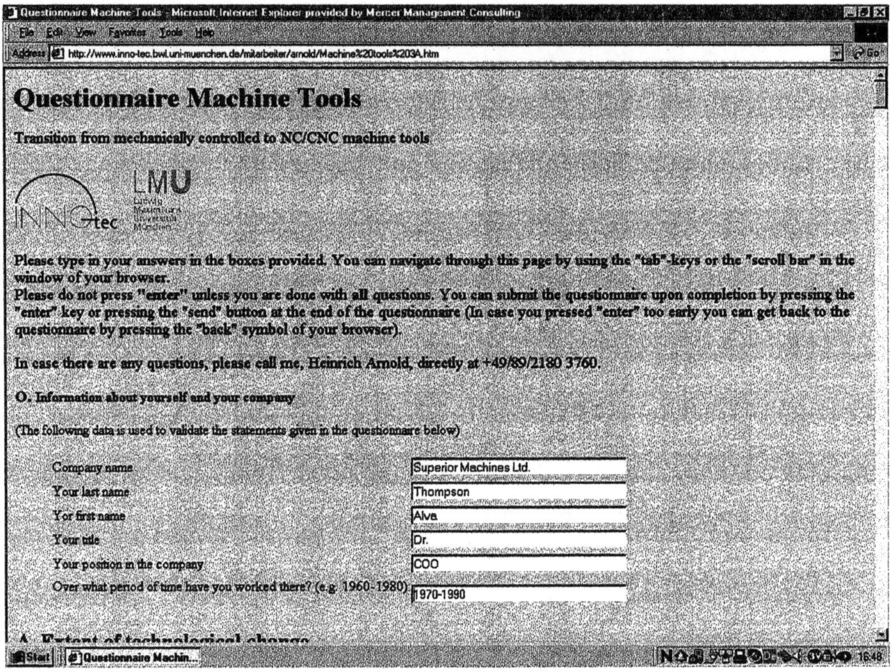

Whether the questionnaire could be opened in a similar way by computers beyond our own domain had to be counter-checked. Sometimes an URL locator seems different for other locations in the web. Thus it had to be ensured that participants - all beyond the university domain - were able to access the questionnaires through the links provided. Similarly, all graphical elements had to be checked for correct presentation as sometimes links to elements in an electronic questionnaire only work within the domain and have to be brought to the same location.

Table A.3.1 below contains the cover letter that preceded the e-mail and on-line questionnaires. The cover letters on paper contained the same information.

Table A.3.1: A typical e-mail cover letter

Dear Mr…
Thank you very much for your assistance.
As I already explained to you briefly over the telephone, I am using the historic advent of NC/CNC technology in the machine tool industry as an example of identifying factors which significantly influence successful management of technology in companies. I am carrying out this study as part of my Ph.D. at the Institute of Innovations Research at the University of Munich, Germany, (see link in the address field below) supported by Prof. Utterback at MIT.

Attached is an abstract of the work I am in the process of doing. I am particularly interested in companies that were either highly successful in incorporating NC/ CNC technology into their portfolio or failed at doing so.
I am using a standardized questionnaire to be able to analyze the data statistically and thus be in a position to identify the decisive factors. Based on an extensive pre-study I came up with a series of standardized questions to be found as a word file in the attachment below or on-line under
http://www.inno-tec.bwl.uni-muenchen.de/mitarbeiter/arnold/Machine%20tools%203A.htm.
Responses will be used purely for statistical purposes. Company specific information will not be published.

Your company is an excellent example of how the success of a company has been influenced by a new technology.

Looking forward to your reply.

With best wishes from Munich,

Heinz Arnold

Engl. Disruptive Technologies ...

Mach tools 2001 Engl. Question...

Heinrich M. Arnold, M.S.Eng., M.B.R.
Munich School of Management
Ludwig-Maximilians-Universität München
Institute for Innovation Research and Technology Management (INNO-tec)
Ludwigstr. 28 RG III
D - 80539 Munich
Tel.: +49 - (0)89 - 2180 3760
Mob.: +49 - (0)170 - 571 4266
Fax: +49 - (0)89 - 2180 6284
arnold@bwl.uni-muenchen.de
http://www.inno-tec.bwl.uni-muenchen.de/mitarbeiter/arnold

Appendix of chapter 4

Table A.4.1: Industry experts in interview series[1]

Mr. Johannes Abler, Managing Director, Liebherr (D)
Mr. Fernando Arocena, Commercial Director, Soraluce – Grupo Danobat (Es)
Mr. Werner Babel, former CEO, MAHO (D)
Mr. Frank Baumbusch, former CTO, Pittler AG (D), Managing Director, Witzig&Frank – Thyssen Krupp Production Systems (D)
Mr. Peter Baumgartner, Director Marketing & Sales, Schneeberger (CH)
Mr. Peter Baumgartner, Managing Director Germany and Head of Manufacturing Group, Mercer Management Consulting (D/ US)
Mr. Guy Bitterolf, Commercial Director, Huron Graffenstaden S.A. (F)
Dr. T. Bouke, Member of Board of Directors, Oerlikon Geartec AG (CH)
Dr.-Ing. Brahms, Division Head Engineering, Ingersoll Waldrich (D/ US)
Mr. Rudi Brandstetter, Head of Engineering Department, Schuler Pressen (D)
Mr. Ernst Braun, Technical Director, Wilhelm Bahmüller (D)
Mr. Dzems Bruvelis, Managing Director and Owner, UBM (D)
Mr. Dieter Bobbert, Managing Director, German Ingersoll Group (D/ US)
Dr. Budde, EXAPT – Verein (D)
Mr. Mario Cuatto, Technical Director, Giana (I)
Mr. Richard Curless, Vice President Product and Technology Development, Cincinnati Machine, a Division of UNOVA Industrial Automation Systems (US)
Mr. Daniel, Technical Director, Bumotec (CH)
Dr. Jan Dannenberg, Principal, Mercer Management Consulting (D/ US)
Mr. Guillermo G. Diaz-Ambrona, Marketing Manager, Nicolas Correa (Es)
Dr. Heinz Drink, Founder and former Managing Director, UBM (D), Vice President, Roland Berger (D), and Mercer Management Consulting (D/ US)
Dr. Walter Eggert, Managing Director, Gleason - Pfauter (US/ D)
Mr. Alessio Facondo, Technical Director, Streparava S.p.A. (I)
Mr. Andi Fischer, Managing Director, Schneeberger (CH)
Dr. Reinhard Fleck, former Director of Development, Production, and Service, Schiess AG (D)
Mr. Klaus Floessel, Managing Director, Schleifring - Ewag AG (D)
Mr. Paul Fricker, former Managing Director, Aciera S.A., Le Locle/NE (CH)
Mr. Wilhelm Gauch, Project- and Sales-Engineer, Burkhardt&Weber (D)
Dr.-Ing. Ö.S. Ganiyusufoglu, Managing Director, Yamazaki Mazak (D/ J)
Mr. Michael Geiger, former Department Head, Ludwigsburger Maschinenbau GmbH (D)
Mr. Bernhard Geis, Economic Department, VDW (D)
Mr. Helmut Goldschmid, former Managing Director, Index (D)
Mr. Frederico Grossrubatscher, former Manager Machine Tool Department, Samputensili (I)
Mr. Alfred Häcker, Managing Director, EMAG (D)
Mr. John Harbough, Head of Marketing, Landis (US)
Mr. Gerhard Hein, Director Marketing, VDW (D)
Mr. Dietmar Hermle, CEO, Hermle AG (D)
Mr. Hans Hofele, Head of Engineering Large Machines, Schuler Pressen (D)

---

[1] Experts and decision makers in italics are not quoted directly in this article but their statements have contributed to and support the overall view.

Mr. Dieter Hofmann, former Managing Director, Heckler&Koch (D)
Mr. Horst Hölzl, former Division Head NC machines, Hüller (D)
Prof. Thomas Parke Hughes, Mellon Professor Emeritus, University of Pennsylvania (US), Distinguished Visiting Professor, Massachusetts Institute of Technology (US)
Mr. Hans Ulrich Jaissle, former CEO, Hüller Hille (D)
Mr. Hiroyuki Kaijima, Managing Director, Toyoda Machinery & Engineering GmbH (D/ J), General Manager, Overseas Sales Department (J)
Mr. Yushiyuki Kanesige, Planning Department, Mori Seiki Co. Ltd. (J)
Dr. Richard L. Kegg, former Vice President of Technology, Milacron, Inc. (US)
Dr. Hans Klein, former Managing Director, Burkhardt&Weber (D)
Prof. Dr. Franz Klenger, former CFO, Gildemeister and Heidenreich&Harbeck (D), now Professor at Fachhochschule Dortmund, Fachbereich Wirtschaft
Prof. Dr.-Ing. Dieter Koerth, former Head of Engineering, Waldrich Siegen – Ingersoll Group (D/ US), now Professor at Fachhochschule Giessen-Friedberg
Mr. Rolf Kölblin, former Managing Director, Honsberg (D)
Mr. Ken Kondo, Head of Export Department, Okuma Corporation (J)
Mr. Peter Klopsch, former Managing Director, Niles Berlin (GDR/ D)
Mr. Paul Kobler, Head of R&D, Starrag (CH)
Mr. Fritz Krempel, former Managing Director, Mauserwerke Oberndorf (D)
Mr. Uwe H. Krüger, President, Chiron Technologies de Production SAS (F)
Mr. Philip Kuo, Overseas Sales Department, Leadwell (Tw)
Mr. Tak Kuratsuwa, Global Marketing Manager, Matsuura Machinery Corporation (J)
Dr. Stephen LeBeau, Product Strategy and Marketing Director, 600 Group (UK)
Mr. Klaus Leinemann, former Principal, Mercer Management Consulting (D/ US)
Mr. Hank Lin, President, Lico Machinery Co., Ltd. (Tw)
Prof. Edward Lumsdaine, ScD, Mechanical Engineering – Engineering Mechanics Dept., Michigan Technological University (US)
Dr. M. Eugene Merchant, Senior Consultant, Institute of Advanced Manufacturing Sciences (US)
Mr. Etore Miletto, Marketing Director, Sicmat (I)
Mr. G. Moillet, COO, Mikron AG (CH)
Mr. H. Nakanishi, International Sales Department, Star Micronics Co. Ltd. (J)
Mr. Rainer Nehls, Vice President, Mercer Management Consulting (D/ US)
Mr. Masayuki Nomura, General Manager, Mori Seiki GmbH (D/ J)
Dr. Gustav Olling, Chairman, Manufacturing Enterprise Council (US), Executive of CAE/CAD/CAM R&D, DaimlerChrysler (US)
Mr. Hironari Okitomo, acting General Manager Engineering Department, Mitsubishi Heavy Industries, Ltd., Machine Tool Division (J)
Dr. Lothar Ophey, former Technical Director, Ex-Cell-O (D/ US)
Mr. Peter Oppelt, Manager Application Engineering, Schleifring – Blohm (D)
Mr. Jean-Robert Passemard, Director Marketing & Products, Renault Automation Comau (F/ I)
Dr.-Ing. Dirk Prust, Division Head, Chiron Werke (D)
Mr. Andreas Quak, Managing Director, Honsberg Lamb – A UNOVA Company (D/ US)
Mr. Ernst Raiser, former Head of Development, Burkhardt&Weber (D)
Dr. Werner Renn, Division Head Large Presses, Müller Weingarten AG (D)
Mr. Werner Rieker, Vice President & General Manager, Giddings&Lewis Machine Tools (US)
Mr. Klaus Rietschel, former Head of Research and Development, Head of Sales, Niles Chemnitz (D), formerly Großdrehmaschinenbau Karl-Marx-Stadt (GDR)
Prof. Dr.-Ing. Hans-Günther Rohs, former Head of Design and Development Division, Boehringer/ Oerlikon Bührle (D/ CH)
Mr. Max Roßkopf, former Head of NC Development, K. Hüller GmbH (D), Managing Director, Team-Technik (D)
Dr. Udo Schapp, Technical Director, Gleason – Hurth (US/ D)
Dr. Klaus Schäfer, former Managing Director, Monforts (D)
Dr. Hans-Martin Schneeberger, CEO, Schneeberger Holding (CH)
Dr.-Ing. Ernst-Friedrich Schimke, former Manager Logistics, Schaudt (D), now Principal and Division Head, Kienbaum Consulting (D)

Mr. Andreas Schinkel, formerly Werkzeugmaschinenfabrik Aschersleben (GDR), now Schiess AG Aschersleben (D)
Mr. Erhard F.J. Schmidt-Staubach, former Managing Director, Niles-Simmons (D)
Mr. Helmut Schneider, former CFO, Ludwigsburger Maschinenbau GmbH Burr (D)
Mr. Robert A. Seach, European Business Development, Hardinge Inc. (B/ US)
Mr. Wolfgang Stöckert, Principal, Mercer Management Consulting (D/ US)
Dr. Zoltan Szabo, former Managing Director, Hüller-Hille GmbH (D)
Mr. Hans Tanner, Managing Director, Schleifring - Studer AG (D/ CH)
Mr. Jack Ting, Assistant Manager, Tong-Tai Machine&Tool Co., Ltd. (Tw)
Prof. Dr. Dr. sc. techn. Hugo Tschirky, ETH Center for Enterprise Science, Group of Technology and Innovation Management (CH)
Prof. Dr. Peter Ulrich, former CEO, Heckert (D)
Mr. Ulrich Vetter, former Managing Director, Schaudt Maschinenbau (D)
Mr. Uve Voumard, CEO, Voumard (CH)
Dr.-Ing. Bernd Weber, Head of Design & Development, Hörmann Industrietechnik GmbH (D)
Mr. Hans-Peter Weber, former Director Machine Tools, Mikron AG Biel (CH)
Prof. Dr. Weidemann, former Head of Marketing, VDF (D), former CEO, Pittler (D)
Prof. Dr. Ulrich Wengenroth, Lehrstuhl für Geschichte der Technik, Technische Universität München (D)
Mr. Wolfgang Weidner, Vice President, Mercer Management Consulting (D/ US)
Dr. Udo Weigel, Vice President Sales, Samputensili (I)
Mr. Lukas Weiss, Director R&D, Tornos Bechler (CH)
Mr. Tetsuro Yoshida, Manager Machine Tool London Head Office, Mitsubishi Heavy Industries Europe, Ltd. (UK/ J)

A.4.1 Categories and components of machine tools and their influence on the industry structure

Machine tools are split up by production process into chip-making and forming machines. The most important making or cutting processes are turning, drilling, boring, milling, and grinding. Important metal forming processes are forging, pressing, bending, and shearing[2]. To date most national statistics follow these categories[3]. Producers' machine portfolios traditionally focus on a single process technology. Before the advent of numerical control technology some producers specialized in turning machines, others in grinding machines, presses, etc. The introduction of numerical controls allowed the combination of several cutting processes into machines with higher flexibility. Likewise most companies increased their knowledge of cutting production technologies. Nevertheless there are until today – with the exception of some large Japanese corporations - only few examples of machine tool producers that manufacture both cutting and forming machine tools.

---

[2] An online reference and explanation of how machine tools work can be found at http://encarta.msn.com/find/Concise.asp?z=1&pg=2&ti=761554377.
[3] As for example in VDW (1999a).

The disruptive effect of the new technology, which was selected, is most visible in chip-making machine tools[4]. This is why we will focus on this group of machine tools when it comes to discussing the technical aspect. The main components of chip-making machine tools are frames, leadways and bearings, drives, and controls (Tschätsch *et al.*, 1991). **Frames** carry mountings for workpieces and tools and support all moving mechanical elements. There are high demands on the rigidity of frames. Undesirable vibrations of the frame result in diminished accuracy in the workpiece.

Producing accurate **leadways and bearings** has been considered an art in the industry of investment goods. Before electronic controls and advanced measurement systems could react 'real time' on feedback from in-process measuring, only very carefully manufactured bearings guaranteed a highly accurate movement of the main spindle, the main spindle driving the tools being the most important mobile part of the machine tool. The accuracy of the workpiece mainly depends on the repetition and positioning accuracy of the spindle with differing numbers of revolutions, drive torques, and cut forces. Equally the accuracy of leadways for other mobile parts (e.g. the table onto which the work piece is mounted) and the quality of mechanical controls influences the precision of the workpiece. Notably Swiss companies, historically closely connected to the watch industry, worked at a high level of precision engineering[5]. The know-how to produce highly precise systems[6], which consists of very accurate leadways, bearings and mechanically controlled drives, was an important competitive advantage acquired over generations, leading to the establishment of a large industrial base for investment goods in Switzerland[7].

Drives consist of electric motors and gears which control the number of revolutions between motor and tool and in addition convert the uniform movement of the motor into a straight, curved, turning or other movement. The main drive is responsible for generating the movements for cutting;

---

4 Dr. Renn, Müller-Weingarten AG, Industry-Expert-Interview-Series (2000-2001).

5 The companies Schneeberger Linear Technologies, Roggwill, and Güdel AG, Langenthal, Switzerland, specialized on linear leadways, are examples of companies which traditionally excel in precision engineering. Their customers are in the machine tool industry but also in the electronics and semiconductor industry, in medical and biotech industries - http://www.schneeberger.com/ and http://www.gudel.com/.

6 The system of these single components has to be highly accurate in combination. I.e. it is not sufficient to produce highly precise components - they also need very careful mechanical integration. See p. 264 of Spur (1996).

7 According to VDW 06/2000 statistics, Switzerland is still number 5 in machine tool production worldwide, after Japan, Germany, USA, and Italy.

additional drives generate thrust and the movements for positioning or shifting gears. Several innovations in particular in electric motors (e.g. step motors, linear motors) have helped to improve drives. The effects of innovations in gear technology were not radical enough to have any effect on the business success of manufacturers of machine tools. Motors were traditionally purchased from suppliers so that the competence[8] of machine tool producers was not deeply affected. Equally, the changes did not have such a radical effect on the whole machine architecture[9]; they were rather modular in character and did not change the product's value proposition[10]. Therefore innovations of drives did not visibly affect the machine tool companies.

The most fundamental changes occurred with the **controls**. Controls switch the drives on and off, they activate the gears and clamps, and direct the movement of the tools. According to which technique they use, controls are classified into mechanical, in some cases into hydraulic and pneumatic, and lastly into electric controls, the latter nowadays usually being based on digital logic. This technology managed to upset the whole industry[11].

A.4.1.1 Numerical control technology

The conventional method of machine tool operation for small lot production (as in contrast to mass production) used mechanical arrangements of bars and cams, as well as sequencing and cycle timers to control the tools of the machines in the production of particular shapes. Fixtures, jigs, and templates, which were often expensive, were used to keep the work in place and to guide the tools[12]. For example the manufacturing of helicopter blades required about fifty templates. The machine tool producers themselves manufactured mechanical as well as hydraulic and pneumatic controls.

---

8 Skills and knowledge as key assets for competitive advantage is discussed in Tushman and Anderson (1986).

9 The extent to which a new technology influences a product's architecture has a significant effect on a company's success or failure. See Henderson and Clark (1990).

10 For a discussion of different dimensions of categorizing technological change and measuring its effect on companies' survival see Arnold (2001).

11 A large part of the above information can be found on pp. 58-122 of Tschätsch and Charchut (1991).

12 For a detailed description of how numerical controls work see p. 137 in Witte (1991).

Digital or numerical[13] controls direct machine tools automatically through a programmed sequence of operations (drilling, milling, boring, etc.) with the aid of computerized control tapes. Machine feed and speed, distance and direction, flow of coolant as well as selection of tools are performed automatically. The change of instructions happens easily by replacing the control tape with another program[14].

Numerical controls marked major changes in the way automatically controlled processes had traditionally been carried out. The new technology was a further step in how information about the workpiece was supplied to the machine. Originally the worker guided the process while he was working with the tools or controlling the process and being present at the machine. In some areas of application, mechanical controls could provide a considerable degree of automation. In particular Swiss screw-type machines were independent of the operator during the production run. The range of shapes was clearly limited but this type of machine was a milestone in the development of an automated process. Digital control technology introduced this level of independence for a larger number of different chip-making machines. NC controls were programmable so the worker did not have to be present during the production process. An automated system for complex parts became possible - the 'automate' (see figure A.4.1 below). Technically this meant the ability to execute real-time control through the use of digital signals corresponding to numbers in contrast to the traditional way of using continuous signals that were analogs of physical quantities.

This change can be compared to the substitution of direct current power by alternating electric power, the emergence of the automobile as a means of transportation at the end of the 19th century or the displacement of vacuum tubes by transistors in the 1960s (Reintjes, 1991).

In the development of numerical controls there have been minor discontinuities (Ehrnberg *et al.*, 1997). The early numerical controls were 'hardwired' and specifically constructed for the various types of machine tools. These first controls lacked flexibility, were expensive and prone to failure. In the early 1970s the numerical control unit was then based on a minicomputer. A Swedish firm initially introduced it. However Japanese firms picked up the computer-based control technology with more momentum and systematically developed it further (Lee, 1996). In 1974 microcomputers were used to make the controls programmable - the computer numerical controls (CNC). This allowed the use of the same control hardware for

[13] The term numerical stems back to the early years of the development project of digital controls at the MIT together with the Parsons Corporation.

[14] The description of functionality follows closely Marx (1979).

a number of machine tool types. Now 'merely' the software had to be modified. With quantities growing, the price for numerical controls dropped and quality stabilized. Implementation of the numerical control technology became simpler for producers and users of machine tools, programming became easier, further automation[15] feasible.

The price/ performance ratio greatly improved and eventually opened up a mass market for numerically controlled machine tools.

Typical for the history of NCs, the initial idea for numerical control was conceived at Parsons Corporation, one of the then leading manufacturers of mechanical parts - the development occurred beyond the scope of the machine tool companies.

Figure A.4.1: Development stages of production processes

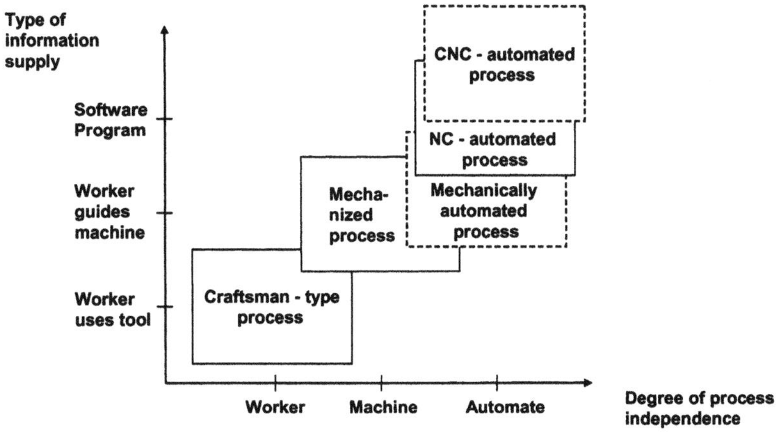

Source: own improved depiction, based on depiction p. 324 of Günter Spur, 'Die Genauigkeit von Maschinen', München, Wien, Carl Hanser Verlag 1996.

15 E.g. for tool changing, material handling, and supervising the production process.

Appendix of chapter 5

A.5.1 Making market share comparable

One of the measurables for the endogenous variable is market share. For this purpose, managers participating in the study were asked to report company sales data in 10-year increments for the years 1960 - 2000. The missing data was supplemented from various sources: statements from interviews with industrial experts[1], consulting reports[2], universities[3], research articles with historic perspective[4], newspaper articles[5], printed company advertising material and histories[6], company histories on the Internet, the AMT Blue Bulletin[7], professional databases[8], and industrial association publications[9]. The original sales data were reported in Deutschmarks, US Dollars, Swiss Franks, Italian Lira, Japanese Yen, French Francs, and Spanish Pesetas and had to be converted into one currency. As exchange rates, the annual means of the daily fixings were used[10]. Drastic changes in

---

1 E.g. Ernst Raiser, former Head of R&D at Burckhardt&Weber. Many thanks to Dzems Bruvelis and Dr. Heinz Drink who enabled access to their personal archives for this study.

2 Mercer-UBM (1992b), Mercer-UBM (1992a), BCG-VDW (1981).

3 Brödner and Schultetus (1992) or Rauleder (1986).

4 Harrop (1985), Forrant (1997).

5 E.g. Handelsblatt excerpts of the 1980, 1981, 1982 editions collected by Dr. Heinz Drink, founder and former managing director of UBM.

6 Kuba (1989), Glunk (1989).

7 Ashburn (1972-2000).

8 This is where we would like to express our gratefulness to Petra Belzer of the agency 'van Dyk', which provided the databases MARCUS and WORLDSCOPE. Mrs. Belzer contributed figures for a number of companies. Furthermore, we would like to thank Ralf Heisig, Information Specialist of Mercer Management Consulting, based in Munich, for his continued advice in establishing contacts with potential sources of information.

9 E.g. USITC (1983) , VDW (1974), VDW (2001), BfA (1979).

10 Time series data 1953/ 1960 to the present day is made available on-line by

currency exchange rates between 1960 and 2000 make it difficult to find comparable market share ratios. Once a company's sales data is compared to world market sales with the aim of calculating a world market share the following problem arises: countries such as Switzerland and Germany experienced a dramatic increase in value of their currencies (factor 2.5 to 3) compared to the US$ between 1960 and 1990.

Although the world market expressed in US$ grew even stronger (by a factor of 12, see table 5.1) during the same period, seen from the perspective of Swiss or German companies, the world market developed at a much slower pace. While a company in the US had to grow by a factor of 12 to maintain its share in the world market, a Swiss or German company had to grow its sales only by a factor of 4. Between 1970 and 1990 the market in nominal DM numbers doubled, whereas expressed in US$ it grew by a factor of more than 5. Inversely, from 1990 to 2000 Swiss and German currencies dropped by 1/6 and 1/5 of their previous exchange rates with the US$ respectively; in dollar value the world market dropped by 1/3. This means that a US company could maintain its market share even if sales dropped by 1/3. A Swiss company, however, could lose only 1/6 in sales in order to maintain its market position, reason being that it also had to balance out the loss in value of the Swiss currency against the US$. Graph A.5.1 shows the development of exchange ratios of currencies of major machine tool producing countries.

Deutsche Bundesbank at http://www.bundesbank.de/de/statistik/zeitreihen/html/ and on-line currency converters at http://stls.frb.org/fred/data/exchange/exjpus and http://cpcug.org/user/invest/datahist.html and others.

Figure A.5.1: Development of currencies

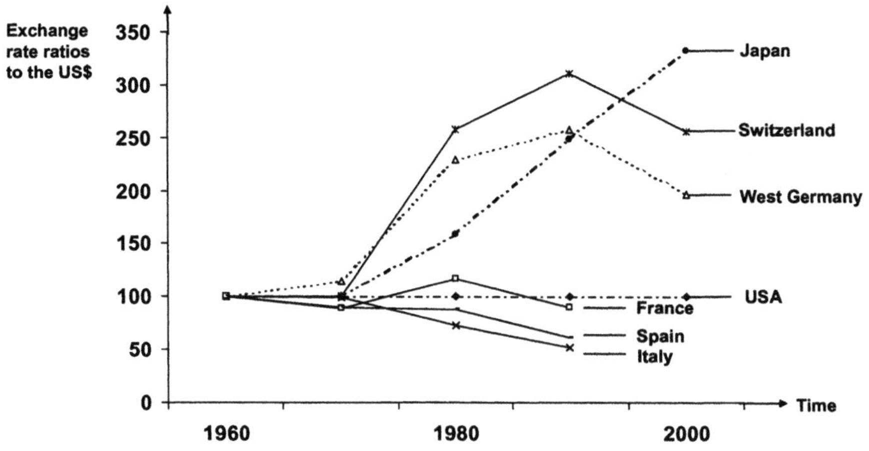

Note: exchange ratio normalized for 1960.
Source: historic archives, own depiction.

Another, although minor aspect, is the geographical location of competitors. Before 1980 competition in the machine tool industry took place mainly on a national basis. The massive entry of Japanese companies accelerated the development of a more global market. Thus, to measure the success of companies one would ideally have to compare them with national competitors up to the 1980s and to international competition thereafter. Taking these two aspects of exchange rate development and geography into account, company sales figures are put in relation with the size of their respective home markets. The figures of consumption (excluding production, import, and export) in the respective home countries of the manufacturers are used as a base of referral for relative competitive success. Introducing this measure for market share success stresses another aspect. It normalizes company sales against the general economic trends within a country and strengthens company-specific market share gains rather than national trends. An example illustrates this: if a company's share related to its country's constant consumption grows, this implies that the company is ahead of its competitors in the domestic market and / or is able to grow sales abroad.

As figure A.5.2 illustrates, national markets develop indeed quite differently. Normalization appears adequate to account for these changes.

Figure A.5.2: Development of domestic markets

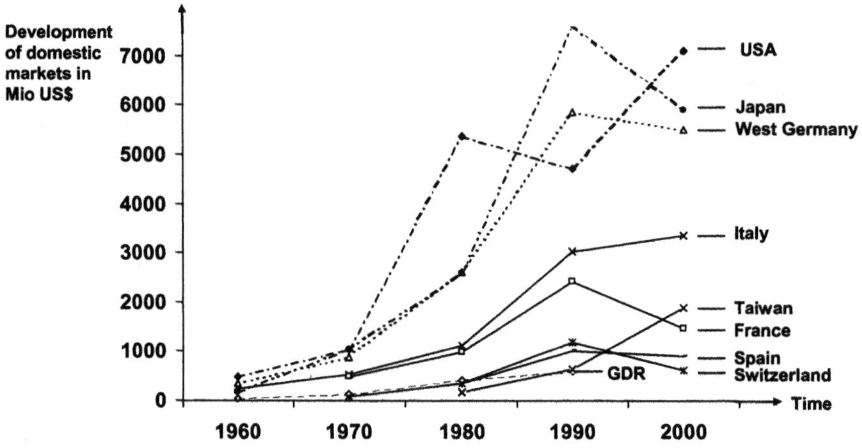

Source: various data sources, own estimates.

Table A.5.1: Questions for variable group 'company attributes'

Variable category and characteristic questions	Shortcut and type	Variable type and span
Ability to forecast market and technology		
▪ Did your company realize **status and trend** of technology timely and accurately compared to the competition?	FORECTEC[1]	1 = too late or wrong; 5 = accurate and timely
▪ What importance did your company attribute to the new market?	Forecmar[1]	1 = no importance; 5 = strategic market
Resistance to innovation		
▪ Was there resistance to the new technology in your company?	RESISTAN[1]	1 = no resistance; 5 = strong resistance
Flexibility of plan		
▪ How flexible was your company's strategic planning process compared to competition?	FLEXBASE[1]	1 = rigid; 5 = highly flexible
▪ **How many times** per year can fundamental **changes** in the strategic plan be made?	FLEXBNUM[2]	Changes per year
▪ **How** flexible was the **detail/subordinate planning processes** in your company compared to competition?	FLEXOP[1]	1 = rigid; 5 = highly flexible
▪ How **many times** per year can incremental **changes** in the **detail/ subordinate plans** be made?	FLEXOPNU[2]	Changes per year
Information flow		
▪ How **formalized** was the information flow **within your company** compared to competition?	INFINTRN[1]	1 = informal; 5 = formalized
▪ How open was the **information exchange** with the **competition**?	INFCOMP[1]	1 = secrecy oriented; 5 = active networking
▪ How open was the **information exchange** with **suppliers**?	INFSUP[1]	1 = secrecy oriented; 5 = active networking
Company culture		
▪ Was **personal initiative/courage** restricted or rewarded sufficiently?	CULTCOUR[1]	1 = restricted; 5 = rewarded sufficiently
▪ Does top management **actively collect new ideas** or does it **rather impede** changes?	CULTCOLL[1]	1 = rather impede changes; 5 = actively collect new ideas
▪ Did **access** to the top management for new ideas exist or were new ideas being **filtered**?	CULTFILT[1]	1 = unfiltered; 5 = highly filtered
Distribution of power		
▪ Who has the **power in your company**: whose ideas get to be implemented?	POWER[1]	1 = lower levels; 5 = executive board

Table A.5.1 (continuation)

Variable category and characteristic questions	Shortcut and type	Variable type and span
Management focus		
▪ What **priority** does **technology management** have in board decisions compared to the competition?	FOCUSTM[1]	1 = low; 5 = strategic
▪ How strongly has **top management** been involved in preparing the decisions on the introduction of NC/CNC compared to competition (e.g. was informed by trade shows, acquisition negotiations, etc.)?	FOCUSINF[1]	1 = not sufficiently; 5 = constantly involved
Consequence of decision making		
▪ How **fast** and **consequent** was decision making and implementation compared to competition?	CONSEQUE[1]	1 = not sufficient; 5 = always on time and binding
Financial base		
▪ How much was spent on **R&D as a percentage** of machine tool **sales** on average 1970 - 1980?	RDPERC[2]	Percentage of Sales
(The RDABS variable is derived from RDPERC and describes the value of average R&D spending in currency)	RDABS[2]	Spending in Mio US$
▪ As a measure of **financial capacity**: What was the average profit of the total company from 1970 - 1980?	RENDABS[2]	Profit in Mio US$
(The RENDPERC variable is derived from RENDABS and describes the average profitability in percentage of sales)	RENDPERC[2]	Percentage of Sales
▪ Total **company revenue** 1975?	SALES75[2]	Revenue in Mio US$
In-house manufacturing		
▪ To what extent could NC/CNC machine tools be **manufactured in-house**?	MANUF[1]	1 = none; 5 = completely
Attractiveness for new hires		
▪ Was your company **able to hire sufficient** personnel to meet goals?	ATTRACT[1]	1 = not sufficient; 5 = sufficient

[1] Self-rating, 5-point interval scale
[2] Metric

Table A.5.2: Questions for variable group 'strategic decisions'

Variable category and characteristic questions	Shortcut and type	Variable type and span
Innotype 1		
What was the **extent of the technological change** from **he mechanically controlled** machine tool to the **first NC nachine tool**; (e.g. has the machine concept been redesigned ompletely, or have existing modules been used)?		
▪ **Incremental** Innovation: The existing **concept** remains unchanged, some **modules** are being **improved**	INNO1INC	Binary[2]
▪ **Modular** Innovation: **Concept** remains **unchanged**, some **modules** are being **replaced/ added on**	INNO1MOD	Binary[2]
▪ **Architectural** Innovation: **New concept of interaction** of modules	INNO1ARC	Binary[2]
▪ **Radical** Innovation: **New strategy of problem solution**	INNO1RAD	Binary[2]
Innotype 2		
What was the **extent of the technological change** from the **first** NC machine tool to **successive NC/CNC machine generations;** has the product architecture been modified again (e.g. complete redesign, or exchange of modules) or have existing modules been improved incrementally?		
▪ **Incremental** Innovation (see above definitions)	INNO2INC	Binary[2]
▪ Modular Innovation	INNO2MOD	Binary[2]
▪ Architectural Innovation	INNO2ARC	Binary[2]
▪ Radical Innovation	INNO2RAD	Binary[2]
Timing of market entry		
When did your company choose to enter the NC/CNC market?		
▪ First to market	MARKT1ST	Binary[2]
▪ Not first to market, but before a dominant design emerged in the industry	MARKT2ND	Binary[2]
▪ After a dominant design has been established	MARKT3RD	Binary[2]
Organizational independence		
What was the organizational approach through which the new technology has been developed and implemented?		
▪ By a separate company	ORGSEP	Binary[2]
▪ By a separate organizational unit (e.g. profit-center, division)	ORGDIV	Binary[2]
▪ From within the existing organization	ORGINTRN	Binary[2]
Business model		
With the advent of the new technology, which of the following business models were formed: a **new** business model (e.g. transition from creating profit through hardware selling to creating profit after sales), a model with **elements of the old business model** (e.g. network effect of installed base), or a model **similar to the old**?		
▪ New business model	BMODNEW	Binary[2]
▪ Contains elements of the old business model	BMODSIM	Binary[2]
▪ Similar to old business model	BMODOLD	Binary[2]

Table A.5.2 (continuation)

Variable category and characteristic questions	Shortcut and type	Variable type and span
Acquisitions		
▪ To what extent have competencies been gained by acquiring companies (total number of employees in the acquired companies)?	TAKEOVER	Number of employees, metric
Alliances & Joint Ventures		
▪ Have there been **alliances** or **joint ventures** to build up competencies?	ALLIANCE	Binary[2]
Changing the value chain		
▪ To what extent was **"old technology"** (e.g. mechanical parts, etc.) being **in-/ or out-sourced** in the course of the technological change?	OUTOLD[1]	1 = complete outsourcing; 5 = complete insourcing
▪ To what extent was **'new technology'** (e.g. controls, software, etc.) being **in-sourced** or **provided by suppliers** in the course of the technological change?	INNEW[1]	1 = completely by suppliers; 5 = complete insourcing

[1] Self-rating, 5-point interval scale
[2] Expressed in dichotomous variable: yes = 1, no = 0.

Table A.5.3: Questions for variable group 'operational decisions'

Variable category and characteristic questions	Shortcut and type	Variable type and span
Resource commitment		
How many **employees** (Empl.) have been devoted to NC/ CNC machines?		
▪ Employees in technical functions 1970 (estimate)	ENGTEC70[2]	Number of employees
▪ Employees in technical functions 2000 (estimate)	ENGTEC00[2]	
▪ Employees in administrative/ business functions 1970 (estimate)	ENGBUS70[2]	
▪ Employees in administrative/ business functions 2000 (estimate)	ENGBUS00[2]	
▪ Employees in technical funct. 1970 as percentage of total	ENGTR70[2]	Percentage
▪ Employees in technical funct. 2000 as percentage of total	ENGTR00[2]	
▪ Employees in business funct. 1970 as percentage of total	ENGTR70[2]	
▪ Employees in business funct. 2000 as percentage of total	ENGTR00[2]	
(The ENGTR and ENGBR variables have been obtained from the ENGTEC and ENBUS variables above)		
▪ What was the ratio of total number of employees at the start of the NC/CNC implementation process to the number of employees before? (express in %)	ORGBEGIN[2]	Percentage
▪ What was the ratio of total number of employees at the end of the digital system implementation process to the number of employees before? (express in %)	ORGEND[2]	Percentage
▪ Growth of part of organization, that deals with NC/ CNC	ORGGROW[2]	Ratio
(ORGGROW is the ratio of ORGEND over ORGBEGIN)		
Level of hierarchy		
▪ How many levels of hierarchy existed in the early organization dealing with the new technology?	HIERLEV[2]	Number of levels
▪ What is the average number of direct subordinates (span of control) in the above-described organization?	HIERBROA[2]	Number of subordinates
Extent of decisions		
▪ What was the **scope** of decisions made related to radical technological change in your company?	DECISION[1]	1 = incremental; 5 = far reaching
Extent of implementation program		
▪ Have **implementation programs** been **radical or incremental** (radical change 'over night' or incremental adaptation)?	REALISAT[1]	1 = radical; 5 = incremental
Cooperation with suppliers and customers		
▪ To what extent was the new **system result of collaboration with suppliers**?	SUPCOOP[1]	1 = hardly; 5 = completely
▪ How strongly was the **value chain** of the new system **controlled by suppliers**?	SUPCONTR[1]	1 = dependent on suppliers; 5 = own control
▪ To what extent was the new **system result of collaboration with customers**?	CLICOOP[1]	1 = hardly; 5 = completely
▪ How strongly was the **value chain** of the new systems **controlled by customers**?	CLICONTR[1]	1 = dependent on customers; 5 = own control

Table A.5.3 (continuation)

Variable category and characteristic questions	Shortcut and type	Variable type and span
Trainings		
To what extent have there been **training programs** due to the technological change?		
▪ **Technical** < 2 weeks (estimate)	TRAINTC2[2]	Number of employees
▪ **Technical** 0,5 - 6 months (estimate)	TRAINTC4[2]	
▪ **Technical** > 6 months (estimate)	TRAINTC6[2]	
▪ **Administrative/ business** < 2 weeks (estimate)	TRAINBS2[2]	
▪ **Administrative/ business** 0,5 - 6 months (estimate)	TRAINBS4[2]	
▪ **Administrative/ business** > 6 months (estimate)	TRAINBS6[2]	
▪ Technical < 2 weeks as percentage of total	TCPRC2[2]	Percentage of total employees 1980
▪ Technical 0,5 - 6 months as percentage of total	TCPRC4[2]	
▪ Technical > 6 months as percentage of total	TCPRC6[2]	
▪ Business < 2 weeks as percentage of total	BSPRC2[2]	
▪ Business 0,5 - 6 months as percentage of total	BSPRC4[2]	
▪ Business > 6 months as percentage of total	BSPRC6[2]	
(The TCPRC and BSPRC variables have been obtained from the TRAINTC and TRAINBS variables above)		
External R&D		
▪ To what extent has your company **purchased external R&D – services** (incl. licenses, patents, contract research)?	RDPURCHA[2]	Accumulated cost
External R&D as percentage of 1980 sales.	RDPURCR[2]	Percentage of 1980 sales
(RDPURCR is the ratio of RDPURCHA and sales of the year 1980)		
▪ To which extent did external services **save in-house development** time?	RDSAVE[2]	Saved development time
▪ Saved development time as percentage of 1980 employees.	RDSAVR[2]	Percentage of total employees 1980
(RDSAVR is the ratio of RDSAVE and total employees in the year 1980)		
New hires		
To what extent were new competencies acquired through **new hires**?		
▪ Employees in **technical functions** (estimate)	TCHIRE[2]	Number of new employees
▪ Employees in **administrative/ business functions** (estimate)	BSHIRE[2]	

1 Self-rating, 5-point interval scale
2 Metric

Comments on the introduction of additional variables

From variables ENGTEC and ENGBUS ratios ENGTR and ENGBR are calculated to express a percentage of total employment. ORGBEGIN and ORGEND, in their original form, are employed in order to determine significant differences, and as a ratio (ORGGROW) to introduce a measure of growing commitment of NC/ CNC implementation from beginning to end.

Unfortunately, there is no adequate distribution available regarding business and technical functions. The training data has thus been changed to ratios of technical training over total number of employees in both technical and business functions. 1980 was chosen as the reference year, as data had been collected from all companies for this year and 1980 deemed a most suitable year to begin with. Indeed, most of the training did take place in the 1980s. The variables generated for these technical functions were TCPERC2, TCPERC4, and TCPERC6, describing the ratio of employees as 'trained less than 2 weeks', 'between 2 weeks and 6 months', and 'more than 6 months', respectively. Similarly, ratios named BSPERC2, BSPERC4, and BSPERC6 were introduced regarding the business functions.

The accumulated cost for external R&D, RDPURCHA, is used together with the companies' sales of the year 1980 to calculate the ratio RDPURCR, which then serves as a comparable inter-company measure for use of external R&D services. Similarly, RDSAVE, resources saved by the use of external R&D, together with the number of companies' total employment in 1980 is used to calculate the ratio RDSAVR.

In case of acquisition of know-how through new hires, the absolute number of employees is used rather than a ratio. We would like to shed light here on the question of how many employees need to be hired to gain critical knowledge.

Table A.5.4: Company attributes - description of total sample (58 companies)

Group and variable	n	Answer range min	Answer range max	Mean	Standard dev.[1] estimator s	Standard error estimator $s/\sqrt{n-1}$
Ability to forecast						
FORECTEC	54	1	5	3,65	1,31	0,18
FORECMAR	53	1	5	3,79	1,23	0,17
Resistance to innovation						
RESISTAN	56	1	5	2,73	1,18	0,16
Flexibility of plan						
FLEXBASE	51	1	5	3,88	1,03	0,14
FLEXBNUM	43	1	12	2,55	3,12	0,48
FLEXOP	48	1	5	3,73	1,05	0,15
FLEXOPNU	40	1	365	51,65	120,06	18,98
Information flow						
INFINTRN	50	1	5	2,84	1,06	0,15
INFCOMP	52	1	5	2,04	1,1	0,15
INFSUP	53	1	5	3,58	1,13	0,16
Company culture						
CULTCOUR	53	2	5	3,74	0,94	0,13
CULTCOLL	52	1	5	3,9	1,11	0,15
CULTFILT	52	1	5	2,37	1,16	0,16
Distribution of power						
POWER	51	3	5	4,02	0,79	0,11
Management focus						
FOCUSTM	51	2	5	4,00	0,94	0,13
FOCUSINV	52	2	5	4,42	0,85	0,12
Consequence of decision making						
CONSEQUE	53	1	5	3,68	1,16	0,16
Financial base						
RDABS	32	0,1	18,8	2,75	3,57	0,63
RDPERC	31	1,1	50	8,73	9,99	1,79
RENDABS	18	0	50	5,61	11,67	2,75
RENDPERC	18	0	31	10,58	8,7	2,05
SALES75	34	0,8	375	46,81	66,34	11,38
In-house manufacturing						
MANUF	55	2	5	3,84	0,92	0,12
Attractiveness for new hires						
ATTRACT	51	1	5	3,63	1,26	0,18

[1] deviation

Table A.5.5: Strategic decisions - description of total sample (58 companies)

Group and variable	n	Answer range min	Answer range max	Mean	Standard dev.[1] estimator s	Standard error estimator $s/\sqrt{n-1}$
Innotype 1						
INNO1INC	53	0	1	0.13	0.34	0.05
INNO1MOD	53	0	1	0.51	0.50	0.07
INNO1ARC	53	0	1	0.19	0.40	0.05
INNO1RAD	53	0	1	0.25	0.43	0.06
Innotype 2						
INNO2INC	54	0	1	0.19	0.39	0.05
INNO2MOD	54	0	1	0.31	0.47	0.06
INNO2ARC	54	0	1	0.39	0.49	0.07
INNO2RAD	54	0	1	0.43	0.50	0.07
Timing of market entry						
MARKT1ST	57	0	1	0.30	0.46	0.06
MARKT2ND	57	0	1	0.44	0.50	0.06
MARKT3RD	57	0	1	0.26	0.44	0.06
Organizational independence						
ORGSEP	55	0	1	0.07	0.26	0.04
ORGDIV	56	0	1	0.09	0.29	0.04
ORGINTRN	55	0	1	0.85	0.36	0.05
Business model						
BMODNEW	52	0	1	0.10	0.30	0.04
BMODSIM	52	0	1	0.28	0.46	0.06
BMODOLD	52	0	1	0.62	0.49	0.07
Acquisitions						
TAKEOVER	40	0.00	2000	147.50	416.94	65.92
Alliances & Joint ventures						
ALLIANCE	51	0	1	0.25	0.44	0.06
Changing the value chain						
OUTOLD	49	1	5	3,08	1,06	0.15
INNEW	49	1	5	2.92	1.13	0.16

[1] deviation

Table A.5.6: Operational decisions - description of total sample (58 companies)

Group and variable	n	Answer range min	Answer range max	Mean	Standard dev.[1] estimator s	Std. error estimator $s/\sqrt{n-1}$
Resource commitment						
ENGTR70	37	0.00	0.70	0.07	0.14	0.02
ENGTR00	40	0.00	0.98	0.40	0.33	0.05
ENGBR70	32	0.00	0.10	0.008	0.02	0.004
ENGBR00	37	0.00	0.36	0.09	0.10	0.02
ORGBEGIN	29	0.00	100.00	21,70	26,26	4,88
ORGEND	30	8.00	120.00	77,73	31,40	5,73
ORGGROW	27	1.00	100	13.44	26.13	5,03
Level of hierarchy						
HIERLEV	41	1	5	2,99	1,02	0.16
HIERBROA	29	2	20	7,31	5,33	0.99
Extent of decisions						
DECISION	48	2	5	0,99	0.99	0.14
Extent of implementation program						
REALISAT	53	1	5	3,60	1,13	0.16
Cooperation with suppliers and customers						
SUPCOOP	55	1	5	3,40	1,16	0.16
SUPCONTR	56	1	5	3,64	1,09	0.15
CLICOOP	55	1	5	3,53	1,00	0.13
CLICONTR	55	0	5	3,51	1,30	0.18
Trainings						
TCPRC2	31	0.00	0.70	0.07	0.14	0.03
TCPRC4	34	0.00	0.60	0.07	0.12	0.02
TCPRC6	33	0.00	0.40	0.02	0.08	0.01
BSPRC2	34	0.00	0.70	0.04	0.12	0.02
BSPRC4	32	0.00	0.80	0.05	0.15	0.03
BSPRC6	33	0.00	0.00	0.00	0.00	0.00
External R&D						
RDPURCHA	34	0.00	31.00	2,33	5,90	1,01
RDPURCR	33	0.00	0.36	0.03	0.07	0.01
RDSAVE	29	0.00	250	16.97	46.61	8,65
RDSAVR	28	0.00	0.83	0.06	0.16	0.03
New hires						
TCHIRE	33	0.00	50.00	12,12	13.77	2,40
BSHIRE	34	0.00	10.00	1,79	3,03	0.52

[1] deviation

Table A.5.7: List of 277 important companies

Number	Company Name	Country	In sample
1	Heid	Austria (A)	
2	Voest-Alpine	A	
3	Aciera S.A.	Switzerland (CH)	✓
4	Agie-Charmilles	CH	
5	Bumotec S.A.	CH	✓
6	Dixi Machines S.A.	CH	
7	Ewag AG	CH	✓
8	Georg Fischer	CH	
9	Hauser	CH	
10	Kellenberger	CH	
11	Klingelnberg-Oerlikon Gruppe, here: OG Tech	CH	✓
12	Kummer Feres	CH	
13	Mägerle AG	CH	
14	Mikron AG, Biel	CH	✓
15	Mikron-Holding	CH	
16	Reiden Technik AG	CH	
17	Reishauer	CH	
18	Schaublin	CH	
19	Starrag	CH	✓
20	Studer AG	CH	✓
21	Tornos-Bechler	CH	✓
22	Voumard	CH	✓
23	Wahli	CH	
24	AEG	Germany/West (D)	
25	Alfing Kessler Maschinenfabrik	D	✓
26	Autania	D	
27	AXA	D	
28	Blohm Maschinenbau GmbH	D	✓
29	Boehringer Maschinen	D	
30	Bohle	D	
31	Boley	D	
32	Bosch	D	
33	Braun	D	
34	Burckhart&Weber	D	✓
35	BWF (Berliner Werkzeugmaschinenfabrik)	D	
36	Carl Benzinger GmbH	D	
37	Carl Hurth (now: Gleason–Hurth)	D	✓
38	Chiron-Werke	D	✓
39	Deckel	D	
40	Diedesheim	D	
41	Dieffenbacher	D	
42	Dörries-Scharmann	D	
43	EMAG Maschinenfabrik GmbH	D	✓
44	Ex-Cell-O GmbH	D	✓
45	Forst	D	
46	Fortuna	D	
47	Fritz Werner	D	
48	Gebr. Hoffmann	D	
49	Gildemeister	D	
50	Gleason-Pfauter	D	✓
51	Grob	D	
52	Grundig	D	

53	H. Kolb	D	
54	Heckler&Koch Maschinen- und Anlagen GmbH	D	✓
55	Heidenhain	D	
56	Heidenreich&Harbeck	D	✓
57	Heilbronn	D	
58	Heller	D	
59	Hermle AG (or: Maschinenfabrik Berthold Hermle	D	✓
60	Hessapp	D	
61	Heyligenstedt	D	
62	Honsberg Lamb	D	✓
63	Hörmann-Barkas Industrietechnik GmbH	D	✓
64	Hüller-Hille	D	✓
65	Index-Werke	D	✓
66	Ingersoll Maschinen und Werkzeuge GmbH – Waldric	D	✓
67	IWKA	D	
68	Jung	D	
69	Junker	D	
70	K. Hüller GmbH	D	✓
71	Krupp	D	
72	Liebherr	D	✓
73	Ludwigsburger Maschinenbau GmbH Burr	D	✓
74	Mägerle AG	D	
75	Maho	D	✓
76	Maschinenfabrik Wilhelm Bahmüller	D	✓
77	Mauserwerke Oberndorf	D	✓
78	Monforts	D	✓
79	Müller Weingarten AG	D	✓
80	Naxos	D	
81	Pittler AG	D	✓
82	Ravensburg	D	
83	Rothenberger	D	
84	Schaudt Maschinenbau	D	✓
85	Schiess AG	D	✓
86	Schleifmaschinenwerk	D	
87	Schleifring	D	
88	Schuler Pressen GmbH & Co. KG	D	✓
89	Schütte	D	
90	SHW	D	
91	Siemens	D	
92	SMG-Pressen	D	
93	Spinner	D	
94	STAMA	D	
95	Steinel	D	
96	Thyssen Maschinenbau	D	(✓)
97	Traub	D	
98	Trumpf	D	
99	Union	D	
100	VDF	D	✓
101	Waldrich Coburg Werkzeugmaschinenfabrik GmbH	D	
102	Walter AG	D	
103	Wanderer	D	
104	Weiler	D	
105	Weisser Söhne	D	
106	WEMA -> Schiess	D	
107	Witzig&Frank	D	
108	Wotan	D	

109	Heckert	Germany/East	✓
110	NILES Chemnitz, vorm. Großmaschinenbau Karl-	DDR	✓
111	Niles Drehmaschinenbau GmbH	DDR	
112	Niles-Simmons	DDR	✓
113	Schiess AG Aschersleben	DDR	✓
114	Umformtechnik Erfurt	DDR	
115	Danobat	Spain (ES)	
116	Ind. Anayak	ES	
117	Jator	ES	
118	Kondia Urbano Conde, S.A.	ES	
119	Lealde	ES	
120	Mupem	ES	
121	Nicolas Correa	ES	✓
122	Soraluce - Grupo Danobat	ES	
123	Tallers	ES	
124	Zayer	ES	
125	Alcera-Gambin	France (F)	
126	Bricaud	F	
127	Cazeneuve	F	
128	Chiron Technologies de Production SAS	F	
129	ETA-Toyoda	F	
130	Huron Graffenstaden	F	
131	Lipemec	F	
132	Manurhin	F	
133	Nodier Emag Industrie	F	
134	Num	F	
135	Realmeca	F	
136	Renault Automation Comau S.A.	F	✓
137	Somab	F	
138	Sonim/ Traub	F	
139	Steinel	F	
140	Wirth et Gruffat	F	
141	Amutio Emag	Italy (I)	
142	Biglia	I	
143	Canavese	I	
144	Castor	I	
145	Comau S.p.A.	I	
146	Cucchi	I	
147	Duplo Standard/ Duplomatic	I	
148	ECS	I	
149	F.M.S.	I	
150	Geminis	I	
151	Giana S.p.A.	I	✓
152	Gioria	I	
153	M.C.M.	I	
154	Maho Graziano	I	
155	Mandelli	I	
156	Meccanica Nova	I	
157	Minganti	I	
158	Novar	I	
159	Olivetti	I	
160	Rambaudi	I	
161	Safop	I	
162	SAIMP/ Pama	I	
163	Samputensili S.p.A.	I	✓
164	Sicmat S.p.A.	I	✓

165	SIGMA S.p.A.	I	
166	Streparava S.p.A.	I	✓
167	Tacchi Giacomo	I	
168	Amada	Japan (J)	
169	Brother/ Sukino	J	
170	Citizen	J	
171	Fanuc	J	
172	Fuji Machine Mfg. Co., Ltd.	J	
173	Hitachi Seiki Co., Ltd.	J	
174	IHI	J	
175	Kitamura	J	
176	Komatsu Ltd.	J	
177	Makino Milling	J	
178	Matsuura	J	
179	Mitsubishi Heavy Industries Europe, Ltd.	J	
180	Miyano	J	
181	Mori Seiki Co. Ltd.	J	✓
182	Nakamura-Tome Precision Industry Co., Ltd.	J	
183	Nippei Toyama	J	
184	Nisshimbo	J	
185	Nomura Automatic Lathe Co., Ltd.	J	
186	Okuma Corp.	J	
187	Star Micronics Co., Ltd.	J	✓
188	Takisawa Machine Tool Co., Ltd.	J	
189	Toshiba Machine Co., Ltd.	J	
190	Toyoda Machine Works	J	✓
191	Tsugami	J	
192	Yamazaki Mazak Corp.	J	✓
193	Yasda	J	
194	Dae Woo Heavy Industries	Korea	
195	Hyundai	Korea	
196	Tong II	Korea	
197	Whacheon	Korea	
198	Philips	Netherlands	
199	Cyclematic Machine Company	Taiwan (TW)	
200	Femco	TW	
201	Leadwell	TW	✓
202	Lico Machinery	TW	✓
203	Proth	TW	
204	Taichung	TW	
205	Tay Yong	TW	
206	Tong-Tai Machine & Tool Co., Ltd.	TW	✓
207	Top-One	TW	
208	Yong Iron	TW	
209	Far East	TW?	
210	Roundtop	TW?	
211	600 Group	United Kingdom	
212	BBC	UK	
213	Beaver	UK	
214	Colchester	UK	
215	Dean Smith & Grace	UK	
216	Denford	UK	
217	Matrix Churchill	UK	
218	Pollard	UK	
219	TS Harrison	UK	
220	Wickmann	UK	

221	Acme Cleveland	United States (US)	
222	American Broach and Machine Co.	US	
223	Bendix	US	
224	Bridgeport Machines Inc.	US	
225	Brown and Sharpe Manufacturing Co.	US	
226	Carlton Machine Tools - NATCO Carlton	US	
227	Chiron USA	US	
228	Cincinnati Milacron	US	✓
229	Colonial Broach	US	
230	Cross&Trecker	US	
231	De Vlieg/ Wadkin	US	
232	Fosdick Machine Tool Co. - Fosdick Kiewert Inc.	US	
233	Gardner	US	
234	GE	US	
235	Giddings&Lewis Machine Tools	US	✓
236	Gleason	US	
237	Hardinge Inc.	US	✓
238	HP	US	
239	Hughes Tool Company	US	
240	Ingersoll Milling	US	
241	Kearney&Trecker	US	
242	Lamb Technicon	US	
243	Landis	US	
244	LaPointe Hudson	US	
245	LaSalle	US	
246	Leadwell/ECO	US	
247	Litton Industries	US	
248	Lodge and Shipley - Monarch Sidney	US	
249	Minster	US	
250	Moline Tool Co.	US	
251	Monarch Machine Tool Co.	US	
252	Moore	US	
253	Morton Manufacturing Co.	US	
254	New Britain Machine Co.	US	
255	Pratt&Whitney	US	
256	Rockford Machine Tool Co.	US	
257	Sheffield Corp.	US	
258	South Bend Lathe Works	US	
259	The Bullard Co.	US	
260	Unova (see Cincinnati Milacron)	US	(✓)
261	Warner&Swasey	US	
262	Waterbug	US	
263	Western Atlas	US	
264	Western Machine Tool Works	US	
265	Cross&Trecker	US	
266	Anilam	Unknown (xy)	
267	DIAG	xy	
268	EBU	xy	
269	ECS	xy	
270	Eumuco	xy	
271	Lauffer	xy	
272	Norte	xy	
273	Redling	xy	
274	Sauter	xy	
275	SMS Hasenclever	xy	
276	Webo	xy	

Table A.5.8: Questions on the importance of company attributes

Variable category and characteristic questions	Shortcut	Observations
Ability to forecast technology		
How would you rate the **importance** of **technology forecasting and evaluation** for the **relative success/ failure** of your company in the industry?	FORECIMP	52
Resistance to innovation		
How would you rate the **importance** of **resistance** for the **relative success/ failure** of your company in the industry?	RESISTIM	51
Flexibility of plan		
How would you rate the **importance** of **planning processes** for the **relative success/ failure** of your company in the industry?	FLEXPIMP	51
Information flow		
How would you rate the **importance** of **internal information flow** for the **relative success/ failure** of your company in the industry?	INFINIMP	51
How would you rate the **importance** of **the external** information flow for the **relative success/ failure** of your company in the industry?	INFEXIMP	51
Company culture		
How would you rate the **importance** of **company culture** for the **relative success/ failure** of your company in the industry?	CULTIMP	52
Distribution of power		
How would you rate the **importance** of **distribution of power** for the **relative success/ failure** of your company in the industry?	POWERIMP	51
Management focus		
How would you rate the **importance** of **management focus** for the **relative success/ failure** of your company in the industry?	FOCUSIMP	52
Consequence of decision making		
How do you rate the **importance** of **timely decision making and consequence of implementation** for the **relative success/ failure** of your company in the industry?	CONSEIMP	53
Financial base		
How would you rate the **importance** of **the financial base** for the **relative success/ failure** of your company in the industry?	RDFINIMP	37
In-house manufacturing		
How would you rate the **importance** of **in-house manufacturing 1970 - 1980** for the **relative success/ failure** of your company in the industry?	MANFOIMP	52
How would you rate the importance of in-house manufacturing **today** for the **relative success/ failure** of your company in the industry?	MANFNIMP	52
Attractiveness for new hires		
How would you rate the **importance of your company's ability to attract new hires** for the **relative success/ failure** of your company in the industry?	ATTRAIM	51

Table A.5.9: Questions on the importance of strategic decisions

Variable category and characteristic questions	Shortcut	Observations
Innotype 1 **How important** was the type of innovation of the **first NC machine tool** for the **relative success/ failure** of your company in the industry?	INNO1IMP	54
Innotype 2 **How important** was the above marked type of innovation of **successive machine concepts** for the **relative success/ failure** of your company in the industry?	INNO2IMP	52
Timing of market entry How would you rate the importance[11] of timing of market entry for the relative success/ failure of your company in the industry?	MARKTIMP	48
Organizational independence How would you rate the **importance** of **organizational independence** for the **relative success/ failure** of your company in the industry?	ORGIMP	52
Business model How would you rate the **importance** of **changing the business model** for the **relative success/ failure** of your company in the industry?	BMODIMP	46
Acquisitions How would you rate the importance of **gaining new competencies by company acquisition** for the **relative success/ failure** of your company in the industry?	TAKEOIMP	38
Alliances & Joint Ventures How would you rate the importance of **alliances and joint ventures** for the **relative success/ failure** of your company in the industry?	ALLIAIMP	49
Changing the value chain How would you rate the importance of **change in the value chain** for the **relative success/ failure** of your company in the industry?	CHAINIMP	48

[11] Importance of this factor compared to other factors below.

Table A.5.10: Questions on the importance of operational decisions

Variable category and characteristic questions	Shortcut	Observations
Resource commitment		
How do you rate the **importance** of **the number of employees devoted to the implementation of NC/CNC** for the **relative success/ failure** of your company in the industry?	ENGIMP	49
Level of hierarchy		
How would you rate the **importance** of **hierarchy and number of subordinates** for the **relative success/ failure** of your company in the industry?	HIERIMP	46
Extent of decisions		
How would you rate the **importance** of the **scope of decision making** for the **relative success/ failure** of your company in the industry?	DECISIMP	49
Extent of implementation program		
How would you rate the importance of the **type of implementation program** for the **relative success/ failure** of your company in the industry?	REALIMP	54
Cooperation with suppliers and customers		
How would you rate the importance of **relations with suppliers** for the **relative success/ failure** of your company in the industry?	SUPPLIMP	53
How would you rate the importance of **relations with customers** for the **relative success/ failure** of your company in the industry?	CLIENIMP	56
Trainings		
How would you rate the importance of **training programs** for the **relative success/ failure** of your company in the industry?	TRAINIMP	41
External R&D		
How would you rate the importance of **purchasing services and competencies** for the **relative success/ failure** of your company in the industry?	RDEXIMP	41
New hires		
How would you rate the importance of **acquiring new competencies through new hires** for the **relative success/ failure** of your company in the industry?	HIREIMP	41

Table A.5.11: Note on t-statistics and modifications used

Z-statistics: $z(X) = \dfrac{\overline{X} - \mu_{pop}}{\sigma/\sqrt{n-1}}$, used for normal distribution in the sample with **known** standard deviation σ of the total population.

T-statistics: $t(X) = \dfrac{\overline{X} - \mu_{pop}}{s/\sqrt{n-1}}$, used for normal distribution in the sample with **unknown** standard deviation σ.

In this case, the population standard error $\sigma/\sqrt{n}$ is replaced by an estimated standard error $s/\sqrt{(n-1)}$. In our case, however, two-sample methods are applied to evaluate the means, based on the assumption that the results of the two samples are statistically independent of each other, both alone as well as compared to each other.

To obtain true t-statistics in our case with the two samples, we assume that the two unknown population variances are the same and have a common variance called homoscedasticity[12]. Test statistics for differences in sample means in this case are:

$$t(X) = \frac{\overline{X}_1 - \overline{X}_2 - D}{s_{pooled}\sqrt{1/n_1 + 1/n_2}}, \text{with}$$

$$D = \mu_1 - \mu_2, \text{and}$$

$$s^2_{pooled} = \frac{(n_1 - 1)s_1^2 + (n_2 - 1)s_2^2}{n_1 + n_2 - 2}$$

To test equality of means, D = 0.

In heteroscedastic cases, where the variances of the two samples are not the same, the standard error is estimated as $s = \sqrt{s_1^2/n_1 + s_2^2/n_2}$. The corresponding test statistics t' are treated like t-statistics, i.e. distributed with a certain degree of freedom 'df'.

[12] P. 414, Hildebrand (1986).

Table A.5.11 (continuation)

$$t'(X) = \frac{\overline{X}_1 - \overline{X}_2}{\sqrt{s_1^2/n_1 + s_2^2/n_2}}, \text{with}$$

$$df = \frac{(n_1 - 1)(n_2 - 1)}{(n_2 - 1)c^2 + (n_1 - 1)(1 - c)^2}, \text{and}$$

$$c = \frac{s_1^2/n_1}{(s_1^2/n_1) + (s_2^2/n_2)}$$

We test for homoscedasticity by means of the Levene test[13]. H_0 is $s_1 \neq s_2$. For $p > 0.05$. There are no statistically relevant differences in the sample group variances.

When variances of the samples are unknown or not the same, the 'Welch-test' is applied[14]. The formula then is as follows:

$$t''(X) = \frac{\left|\overline{X}_1 - \overline{X}_2\right|}{\sqrt{s_1^2/n_1 + s_2^2/n_2}}, \text{with}$$

$$df = \frac{\left(s_1^2/n_1 + s_2^2/n_2\right)^2}{(n_2 - 1)c^2 + (n_1 - 1)(1 - c)^2}, \text{and}$$

$$c = \frac{s_1^2/n_1}{(s_1^2/n_1) + (s_2^2/n_2)}$$

---

[13] As implemented in standard statistics software such as SPSS. See p. 223 Bühl and Zöfel (2000).

[14] See p. 145 in Toutenburg (2000).

Table A.5.12'a': Company attributes - comparison of winners and losers of all shocks (waves 1 - 3, all product types)

Group and variable	'All' Mean	'Losers' n_1	Mean$_1$	$s_1/\sqrt{n_1-1}$	'Winners' n_2	Mean$_2$	$s_2/\sqrt{n_2-1}$	Difference of Means[1]	Significance level[2] p
Ability to forecast technology									
FORECTEC	3,65	32	3,56	0.24	22	3,77	0.27	0.21	n.s.
FORECMAR K-S Z-TEST	3,79	32	3,59	0.23	21	4,1	0.23	0.5	n.s. (0.148)
Resistance to innovation									
RESISTAN	2,73	33	2,82	0.19	23	2,61	0.27	-0.21	n.s.
Flexibility of plan									
FLEXBASE K-S Z-TEST	3,88	30	3,83	0.17	21	3,95	0.26	1.2	n.s.
FLEXBNUM K-S Z-TEST	2,55	25	2.00	0.44	18	3,31	0.95	1.31	n.s.
FLEXOP	3,73	28	3,75	0.18	20	3,7	0.27	-0.05	n.s.
FLEXOPNU K-S Z-TEST	51.65	24	51.88	24.70	16	51.31	30.65	-0.56	n.s.
Information flow									
INFINTRN	2,84	29	2,79	0.17	21	3,91	0.28	0.11	n.s.
INFCOMP	2,04	30	3,43	0.18	22	4,55	0.28	-0.15	n.s.
INFSUP K-S Z-TEST	3,58	31	3,52	0.20	22	3,68	0.26	0.17	n.s.
Company culture									
CULTCOUR K-S Z-TEST	3,74	30	3,6	0.14	23	3,91	0.23	0.31	n.s.
CULTCOLL K-S Z-TEST	3,9	30	3,43	0.21	22	4,55	0.14	1.11***	0.006 (0.000)
CULTFILT K-S Z-TEST	2,37	30	2,67	0.18	22	1,95	0.28	-0.71**	0.039 (0.027)
Distribution of power									
POWER	4,02	29	4,07	0.15	22	3,95	0.17	-0.11	n.s.
Management focus									
FOCUSTM K-S Z-TEST	4.00	30	3,7	0.17	21	4,43	0.16	0.73+	0.119 (0.005)
FOCUSINV K-S Z-TEST	4,42	30	4,2	0.18	22	4,73	0.10	0.53	n.s. (0.014)
Consequence of decision making									
CONSEQUE	3,68	30	3,47	0.21	23	35125	0.24	0.49+	0.128
Financial base									
RDABS	2,75	18	3,04	0.99	14	2,38	0.72	-0.67	n.s.
RDPERC	8,73	18	6,95	0.92	13	11,2	4,08	4.25	n.s.
RENDABS	5,61	10	7,74	4,86	8	2,95	1,26	-4.79	n.s.
RENDPERC	10,58	10	9,44	2,8	8	12.00	3,13	2.57	n.s.
SALES75 K-S Z-TEST	46.81	21	49.67	17.13	13	42.20	11,76	-7.46	n.s.
In-house manufacturing									
MANUF K-S Z-TEST	3,84	32	3,97	0.14	23	3,65	0.22	-0.32	n.s.
Attractivity for new hires									
ATTRACT	3,63	30	3,67	0.21	21	3,57	0.32	-0.1	n.s.

[1] Average difference in mean values. [2] Significance levels of differences in mean values: *** <= 0.01, ** <= 0.05, * <= 0.1, + <= 0.15, n.s. = not significant.
K-S Z-test Variables tested by means of the Kolmogorov-Smirnov Z-test in case of no evidence of normal distribution (significance levels of t-tests in brackets).
All other variables tested by means of the t-test and checked with the Levene-Test for variance equality.

Table A.5.13'a': Company attributes - comparison of winners and losers of early shocks (waves 1 and 2, all product types)

Group and variable	'All'	'Losers'			'Winners'			Difference of Means[1]	Significance level[2]
	Mean	n_1	$Mean_1$	$s_1/\sqrt{n_1-1}$	n_2	$Mean_2$			p
Ability to forecast technology									
FORECTEC [K-S Z-TEST]	3,65	15,00	3,33	0.41	37,00	3,78	0.20	0.45	n.s.
FORECMAR [K-S Z-TEST]	4,78	15,00	3,07	0.33	36,00	4,08	0.18	1.02*	0.051 (0.006)
Resistance to innovation									
RESISTAN	2,74	15,00	3,20	0.30	39,00	2,56	0.19	-0.64*	0.081
Flexibility of plan									
FLEXBASE [K-S Z-TEST]	3,90	15,00	3,67	0.23	34,00	4,00	0.19	0.33	n.s.
FLEXBNUM [K-S Z-TEST]	2,62	12,00	2,33	0.90	29,00	2,74	0.60	0.41	n.s.
FLEXOP	3,74	14,00	3,71	0.24	32,00	3,75	0.20	0.04	n.s.
FLEXOPNU [K-S Z-TEST]	54.29	12,00	7,75	1,75	26,00	75,77	28.24	68.02	n.s. (0.024)
Information flow									
INFINTRN	2,83	15,00	2,93	0.21	33,00	2,79	0.20	-0.15	n.s.
INFCOMP [K-S Z-TEST]	2,04	15,00	2,20	0.26	35,00	1,97	0.20	-0.23	n.s.
INFSUP [K-S Z-TEST]	3,61	15,00	3,33	0.32	36,00	3,72	0.18	0.39	n.s.
Company culture									
CULTCOUR	3,75	15,00	3,67	0.21	36,00	3,78	0.17	0.11	n.s.
CULTCOLL [K-S Z-TEST]	3,92	15,00	3,33	0.30	35,00	4,17	0.17	0.84	n.s. (0.014)
CULTFILT [K-S Z-TEST]	2,32	15,00	2,80	0.24	35,00	2,11	0.20	-0.69+	0.128 (0.052)
Distribution of power									
POWER	4,02	15,00	4,00	0.22	34,00	4,03	0.14	0.03	n.s.
Management focus									
FOCUSTM [K-S Z-TEST]	4,00	15,00	3,63	0.22	34,00	4,21	0.16	0.67+	0.122 (0.022)
FOCUSINV [K-S Z-TEST]	4,02	15,00	4,20	0.22	35,00	4,51	0.14	0.31	n.s
Consequence of decision making									
CONSEQUE	3,65	15,00	3,20	0.30	36,00	3,83	0.19	0.63*	0.076
Financial base									
RDABS	2,84	9,00	1,80	0.56	22,00	3,26	0.87	1.46	n.s.
RDPERC [K-S Z-TEST]	8,90	9,00	5,91	1,43	21,00	10,18	2,54	4.27	n.s.
RENDABS	5,92	6,00	3,72	2,11	11,00	7,12	4,38	3.40	n.s.
RENDPERC	10,32	6,00	11,57	4,36	11,00	9,64	8,28	-1.94	n.s.
SALES75	48,15	10,00	33,09	7,92	23,00	54,70	16.29	21.60	n.s.
In-house manufacturing									
MANUF [K-S Z-TEST]	3,85	15,00	4,00	0.17	38,00	3,79	0.16	-0.21	n.s.
Attractiveness for new hires									
ATTRACT [K-S Z-TEST]	3,63	15,00	3,80	0.28	34,00	3,56	0.24	-0.24	n.s.

[1] Average difference in mean values. [2] Significance levels of differences in mean values: *** <= 0.01, ** <= 0.05, * <= 0.1, + <= 0.15, n.s. = not significant.
[K-S Z-test] *Variables tested by means of the Kolmogorov-Smirnov Z-test in case of no evidence of normal distribution* (significance levels of t-tests in brackets).
All other variables tested by means of the t-test and checked with the Levene-Test for variance equality.

Table A.5.14'a': Company attributes - comparison of winners and losers of recent shock (wave 3, all product types)

Group and variable	'All'	'Losers'			'Winners'			Difference of Means[1]	Significance level[2] p
	Mean	n_1	$Mean_1$	$s_1/\sqrt{n_1-1}$	n_2	$Mean_2$	$s_2/\sqrt{n_2-1}$		
Ability to forecast technology									
FORECTEC [K-S Z-test]	3.67	24	3.46	0.27	25	3.88	0.25	0.42	n.s.
FORECMAR	3.92	24	3.75	0.28	24	4.08	0.21	0.33	n.s
Resistance to innovation									
RESISTAN	2.59	25	2.56	0.19	26	2.62	0.25	0.06	n.s.
Flexibility of plan									
FLEXBASE [K-S Z-test]	3.89	22	3.86	0.20	24	3.92	0.24	0.05	n.s.
FLEXBNUM [K-S Z-test]	2.64	20	2.10	0.54	20	3.18	0.86	1.07	n.s.
FLEXOP	3.70	21	3.67	0.21	23	3.74	0.25	0.07	n.s.
FLEXOPNU [K-S Z-test]	55.65	19	63.91	30.75	18	47.00	27.31	-16.84	n.s.
Information flow									
INFINTRN	2.87	21	2.90	0.21	24	2.83	0.25	-0.07	n.s.
INFCOMP [K-S Z-test]	1.94	22	1.95	0.19	25	1.92	0.24	-0.03	n.s.
INFSUP [K-S Z-test]	3.58	23	3.48	0.25	25	3.68	0.23	0.20	n.s.
Company culture									
CULTCOUR [K-S Z-test]	3.73	22	3.55	0.14	26	3.88	0.22	0.34	n.s.
CULTCOLL [K-S Z-test]	4.00	22	3.45	0.26	25	4.48	0.14	1.03***	0.006 (0.002)
CULTFILT [K-S Z-test]	2.28	22	2.59	0.20	25	2.00	0.26	-0.59**	0.039 (0.085)
Distribution of power									
POWER	4.07	21	4.19	0.18	25	3.96	0.16	-0.23	n.s.
Management focus									
FOCUSTM [K-S Z-test]	4.04	22	3.68	0.22	24	4.38	0.16	0.69+	0.119 (0.013)
FOCUSINV [K-S Z-test]	4.49	22	4.27	0.22	25	4.68	0.11	0.41	n.s. (0.109)
Consequence of decision making									
CONSEQUE	3.65	22	3.41	0.25	26	3.85	0.23	0.44	n.s
Financial base									
RDABS	2.93	13	3.90	1.29	16	2.14	0.65	-1.77	n.s.
RDPERC	9.24	13	7.57	0.97	15	10.69	3.58	3.11	n.s.
RENDABS	5.75	7	10.48	6.80	10	2.44	1.05	-8.04	n.s.
RENDPERC	10.61	7	10.15	3.78	10	10.94	2.76	0.78	n.s.
SALES75 [K-S Z-test]	49.15	16	60.01	21.94	15	37.58	10.61	-22.43	n.s.
In-house manufacturing									
MANUF [K-S Z-test]	3.80	24	3.96	0.18	26	3.65	0.20	-0.30	n.s.
Attractiveness for new hires									
ATTRACT	3.59	22	3.64	0.22	24	3.54	0.29	-0.09	n.s.

[1] Average difference in mean values. [2] Significance levels of differences in mean values: ***<= 0.01, **<= 0.05, * <= 0.1, + <= 0.15, n.s. = not significant.
[K-S Z-test] Variables tested by means of the Kolmogorov-Smirnov Z-test in case of no evidence of normal distribution (significance levels of t-tests in brackets).
All other variables tested by means of the t-test and checked with the Levene-Test for variance equality.

Table A.5.15'p': Company attributes - comparison of winners and losers of all shocks (waves 1 - 3, without presses)

Group and variable	'All' Mean	'Losers' n_1	'Losers' $Mean_1$	'Losers' $s_1/\sqrt{n_1-1}$	'Winners' n_2	'Winners' $Mean_1$	'Winners' $s_2/\sqrt{n_2-1}$	Difference of Means[1]	Significance level[2] p
Ability to forecast technology									
FORECTEC K-S Z-test	3.64	32	3.56	0.24	21	3.76	0.28	0.20	n.s.
FORECMAR	3.79	32	3.59	0.23	20	4.10	0.24	0.51+	0.136
Resistance to innovation									
RESISTAN	2.73	33	2.82	0.19	22	2.59	0.28	-0.23	n.s.
Flexibility of plan									
FLEXBASE K-S Z-test	3.88	30	3.83	0.17	21	3.95	0.26	0.12	n.s.
FLEXBNUM K-S Z-test	2.56	25	2.00	0.44	17	3.38	1.01	1.38	n.s.
FLEXOP	3.79	28	3.75	0.18	19	3.84	0.24	0.09	n.s.
FLEXOPNU K-S Z-test	52.85	24	51.88	24.70	15	54.40	32.60	2.52	n.s.
Information flow									
INFINTRN	2.86	29	2.79	0.17	20	2.95	0.29	0.16	n.s.
INFCOMP	2.06	30	2.10	0.18	21	2.00	0.28	-0.10	n.s.
INFSUP K-S Z-test	3.58	31	3.52	0.20	21	3.67	0.27	0.15	n.s.
Company culture									
CULTCOUR K-S Z-test	3.75	30	3.60	0.14	22	3.95	0.24	0.35	n.s.
CULTCOLL K-S Z-test	3.90	30	3.43	0.21	21	4.57	0.15	1.14***	0.008 (0.000)
CULTFILT K-S Z-test	2.35	30	2.67	0.18	21	1.90	0.28	-0.76**	0.020 (0.020)
Distribution of power									
POWER	4.02	29	4.07	0.15	21	3.95	0.18	-0.12	n.s.
Management focus									
FOCUSTM K-S Z-test	4.00	30	3.70	0.17	20	4.45	0.17	0.75*	0.079 (0.005)
FOCUSINV K-S Z-test	4.43	30	4.20	0.18	21	4.76	0.10	0.56***	0.009
Consequence of decision making									
CONSEQUE	3.69	30	3.47	0.21	22	4.00	0.25	0.53+	0.103
Financial base									
RDABS	2.71	18	3.04	0.99	13	2.25	0.77	-0.79	n.s.
RDPERC	8.88	18	6.95	0.92	12	11.77	4.39	4.82	n.s.
RENDABS	5.66	10	7.74	4.86	7	2.68	1.42	-5.06	n.s.
RENDPERC	10.90	10	9.44	2.81	7	12.98	3.44	3.54	n.s.
SALES75 K-S Z-test	45.40	21	49.67	17.13	12	37.94	11.91	-11.73	n.s.
In-house manufacturing									
MANUF K-S Z-test	3.83	32	3.97	0.14	22	3.64	0.23	-0.33	n.s.
Attractiveness for new hires									
ATTRACT	3.62	30	3.67	0.21	20	3.55	0.34	-0.12	n.s.

[1] Average difference in mean values. [2] Significance levels of differences in mean values: *** <= 0.01, **< = 0.05, *< = 0.1, + <= 0.15, n.s. = not significant.
K-S Z-test Variables tested by means of the Kolmogorov-Smirnov Z-test in case of no evidence of normal distribution (significance levels of t-tests in brackets).
All other variables tested by means of the t-test and checked with the Levene-Test for variance equality.

Table A.5.16'p': Company attributes - comparison of winners and losers of early shocks (waves 1 and 2, without presses)

Group and variable	'All'	'Losers'			'Winners'			Difference	Significance
	Mean	n_1	$Mean_1$	$s_1/\sqrt{n_1-1}$	n_2	$Mean_1$	$s_2/\sqrt{n_2-1}$	of Means[1]	level[2] p
Ability to forecast technology									
FORECTEC [K-S Z-test]	3.65	15	3.33	0.41	36	3.78	0.20	0.44	n.s.
FORECMAR [K-S Z-test]	3.78	15	3.07	0.33	35	4.09	0.19	1.02*	0.059 (0.007)
Resistance to innovation									
RESISTAN	2.74	15	3.20	0.30	38	2.55	0.19	-0.65*	0.080
Flexibility of plan									
FLEXBASE [K-S Z-test]	3.90	15	3.67	0.23	34	4.00	0.19	0.33	n.s.
FLEXBNUM [K-S Z-test]	2.64	12	2.33	0.90	28	2.77	0.63	0.43	n.s.
FLEXOP	3.80	14	3.71	0.24	31	3.84	0.19	0.13	n.s.
FLEXOPNU [K-S Z-test]	55.62	12	7.75	1.75	25	78.60	29.25	70.85	n.s. (0.024)
Information flow									
INFINTRN	2.85	15	2.93	0.21	32	2.81	0.21	-0.12	n.s.
INFCOMP [K-S Z-test]	2.06	15	2.20	0.26	34	2.00	0.20	-0.20	n.s.
INFSUP [K-S Z-test]	3.60	15	3.33	0.32	35	3.71	0.19	0.38	n.s.
Company culture									
CULTCOUR [K-S Z-test]	3.76	15	3.67	0.21	35	3.80	0.17	0.13	n.s.
CULTCOLL [K-S Z-test]	3.92	15	3.33	0.30	34	4.18	0.18	0.84	n.s. (0.015)
CULTFILT [K-S Z-test]	2.31	15	2.80	0.24	34	2.09	0.20	-0.71*	0.098 (0.046)
Distribution of power									
POWER	4.02	15	4.00	0.22	33	4.03	0.14	0.03	n.s.
Management focus									
FOCUSTM [K-S Z-test]	4.00	15	3.53	0.22	33	4.21	0.17	0.68*	0.099 (0.023)
FOCUSINF [K-S Z-test]	4.43	15	4.20	0.22	34	4.53	0.15	0.33	n.s
Consequence of decision making									
CONSEQUE	3.66	15	3.20	0.30	35	3.86	0.19	0.66*	0.069
Financial base									
RDABS	2.80	9	1.80	0.56	21	3.22	0.92	1.42	n.s.
RDPERC [K-S Z-test]	9.05	9	5.91	1.43	20	11.86	2.65	4.56	n.s.
RENDABS	5.99	6	3.72	2.11	10	7.35	4.83	3.63	n.s.
RENDPERC	10.64	6	11.57	4.35	10	10.08	2.72	0.76	n.s.
SALES75 [K-S Z-test]	3.85	10	33.09	7.93	22	52.94	16.95	19.85	n.s.
In-house manufacturing									
MANUF [K-S Z-test]	3.85	15	4.00	0.17	38	3.79	0.16	-0.22	n.s
Attractiveness for new hires									
ATTRACT [K-S Z-test]	3.62	15	3.80	0.28	34	3.56	0.24	-0.25	n.s.

[1] Average difference in mean values. [2] Significance levels of differences in mean values: *** <= 0.01, ** <= 0.05, * <= 0.1, + <= 0.15, n.s. = not significant.
[K-S Z-test] Variables tested by means of the Kolmogorov-Smirnov Z-test in case of no evidence of normal distribution (significance levels of t-tests in brackets).
All other variables tested by means of the t-test and checked with the Levene-Test for variance equality.

Table A.5.17'p': Company attributes - comparison of winners and losers of recent shock (wave 3, without presses)

Group and variable	'All'	'Losers'			'Winners'			Difference	Significance
	Mean	n_1	$Mean_1$	$s_1/\sqrt{n_1-1}$	n_2	$Mean_1$	$s_2/\sqrt{n_2-1}$	of Means[1]	level[2] p
Ability to forecast technology									
FORECTEC K-S Z-test	3.67	24	3.46	0.27	24	3.88	0.26	0.42	n.s.
FORECMAR	3.91	24	3.75	0.28	23	4.09	0.22	0.34	n.s
Resistance to innovation									
RESISTAN	2.58	25	2.56	0.19	25	2.60	0.26	0.04	n.s.
Flexibility of plan									
FLEXBASE K-S Z-test	3.89	22	3.86	0.20	24	3.92	0.24	0.05	n.s.
FLEXBNUM K-S Z-test	2.65	20	2.10	0.54	19	3.24	0.91	1.14	n.s.
FLEXOP	3.77	21	3.67	0.21	22	3.86	0.22	0.20	n.s.
FLEXOPNU K-S Z-test	57.06	19	63.91	30.75	17	49.47	28.85	-14.37	n.s.
Information flow									
INFINTRN	2.89	21	2.90	0.21	23	2.87	0.25	-0.03	n.s.
INFCOMP K-S Z-test	1.96	22	1.95	0.19	24	1.96	0.25	0.01	n.s.
INFSUP K-S Z-test	3.57	23	3.48	0.25	24	3.67	0.24	0.19	n.s.
Company culture									
CULTCOUR K-S Z-test	3.74	22	3.55	0.14	25	3.92	0.23	0.37+	0.106 (n.s.)
CULTCOLL K-S Z-test	4.00	22	3.45	0.26	24	4.50	0.15	1.05**	0.035 (0.050)
CULTFILT K-S Z-test	2.26	22	2.59	0.20	24	1.96	0.27	-0.63*	0.065 (0.069)
Distribution of power									
POWER	4.07	21	4.19	0.18	24	3.96	0.16	-0.23	n.s.
Management focus									
FOCUSTM K-S Z-test	4.04	22	3.68	0.22	23	4.39	0.16	0.71	n.s. (0.013)
FOCUSINV K-S Z-test	4.50	22	4.27	0.22	24	4.71	0.11	0.44	n.s. (0.088)
Consequence of decision making									
CONSEQUE	3.66	22	3.41	0.25	25	3.88	0.23	0.47	n.s
Financial base									
RDABS	2.89	13	3.91	1.29	15	2.01	0.68	-1.90	n.s.
RDPERC	9.42	13	7.58	0.93	14	11.14	3.82	3.56	n.s.
RENDABS	5.80	7	10.48	6.81	9	2.17	1.14	-8.31	n.s.
RENDPERC	10.95	7	10.15	3.78	9	11.57	3.00	1.42	n.s.
SALES75 K-S Z-test	47.68	16	60.00	21.94	14	33.59	10.56	-26.41+	0.102 (n.s.)
In-house manufacturing									
MANUF K-S Z-test	3.80	24	3.96	0.18	25	3.64	0.21	-0.32	n.s.
Attractiveness for new hires									
ATTRACT	3.58	22	3.64	0.26	23	3.52	0.30	-0.11	n.s.

[1] Average difference in mean values. [2] Significance levels of differences in mean values: ***<=0.01, **<=0.05, *<=0.1, +<=0.15, n.s. = not significant.

K-S Z-test Variables tested by means of the Kolmogorov-Smirnov Z-test in case of no evidence of normal distribution (significance levels of t-tests in brackets).

All other variables tested by means of the t-test and checked with the Levene-Test for variance equality.

Table A.5.18'a': Strategic decisions - comparison of winners and losers of all shocks (waves 1 - 3, all product types)

Group and variable	'All' Mean	'Losers' R. freq.$_1$[3]	'Losers' $Mean_1$	'Losers' $s_1/\sqrt{n_1-1}$	'Winners' R. freq.$_2$[3]	'Winners' $Mean_2$	'Winners' $s_2/\sqrt{n_2-1}$	Difference of Means[1]	Significance level[2] p
Innotype 1									
INNO1INC	0.13	2/30	0.07	0.05	5/23	0.22	0.08	0.15	n.s.
INNO1MOD	0.51	20/30	0.67	0.09	7/23	0.30	0.10	-0.36*	0.066
INNO1ARC	0.19	5/30	0.17	0.07	5/23	0.22	0.09	0.05	n.s.
INNO1RAD	0.25	5/30	0.17	0.07	8/23	0.35	0.10	0.18	n.s.
Innotype 2									
INNO2INC	0.19	34/31	0.13	0.06	6/23	0.26	0.09	0.13	n.s.
INNO2MOD	0.31	5/31	0.16	0.07	12/23	0.52	0.11	0.36*	0.065
INNO2ARC	0.39	17/31	0.55	0.09	4/23	0.17	0.08	-0.37**	0.049
INNO2RAD	0.43	14/31	0.45	0.09	9/23	0.39	0.10	-0.06	n.s.
Timing of market entry									
MARKT1ST	0.30	11/32	0.34	0.09	6/25	0.24	0.09	-0.10	n.s.
MARKT2ND	0.44	18/32	0.56	0.09	7/25	0.28	0.09	-0.28	n.s.
MARKT3RD	0.26	3/32	0.09	0.05	12/25	0.48	0.10	0.39**	0.030
Organizational independence									
ORGSEP	0.07	2/32	0.06	0.04	2/23	0.09	0.06	0.02	n.s.
ORGDIV	0.09	3/32	0.09	0.05	2/24	0.08	0.06	-0.01	n.s.
ORGINTRN	0.85	28/32	0.88	0.06	19/23	0.83	0.08	-0.05	n.s.
Business model									
BMODNEW	0.10	2/29	0.07	0.05	3/23	0.13	0.07	0.06	n.s.
BMODSIM	0.29	4/29	0.14	0.07	11/23	0.47	0.11	0.34+	0.102
BMODOLD	0.62	23/29	0.79	0.08	9/23	0.39	0.10	-0.40**	0.032
Acquisitions									
TAKEOVER $^{K\text{-}S\ Z\text{-}test}$	147.50	26[4]	165.38	94.45	14[4]	114.29	72.52	-51.09	n.s.
Alliances & Joint ventures									
ALLIANCE	0.25	8/30	0.27	0.08	5/21	0.24	0.10	-0.03	n.s.
Changing the value chain									
OUTOLD $^{K\text{-}S\ Z\text{-}test}$	3.08	29[4]	3.00	0.17	20[4]	3.20	0.28	0.20	n.s.
INNEW $^{K\text{-}S\ Z\text{-}test}$	2.92	29[4]	2.72	0.19	20[4]	3.20	0.28	0.48	n.s.

[1] Average difference in mean values. [2] Significance levels of differences in mean values: ***<= 0.01, **<= 0.05, * <= 0.1, + <= 0.15, n.s. = not significant.
[3] Relative frequency. [4] Quantities n_1 and n_2 for interval-scaled parameters.
$^{K\text{-}S\ Z\text{-}test}$ Variables tested by means of the Kolmogorov-Smirnov Z-test in case of no evidence of normal distribution (significance levels of t-tests in brackets).
All other variables tested by means of the t-test and checked with the Levene-Test for variance equality.

Table A.5.19'a': Strategic decisions - comparison of winners and losers of early shocks (Waves 1 and 2, all product types)

Group and variable	'All' Mean	'Losers' R. freq.$_1$[3]	'Losers' Mean$_1$	'Losers' $s_1/\sqrt{n_1-1}$	'Winners' R. freq.$_2$[3]	'Winners' Mean$_2$	'Winners' $s_2/\sqrt{n_2-1}$	Difference of Means[1]	Significance level[2] p
Innotype 1									
INNO1INC	0.14	0/15	0.00	0.00	7/39	0.18	0.06	0.18	n.s.
INNO1MOD	0.51	6/12	0.50	0.15	20/39	0.51	0.08	0.01	n.s.
INNO1ARC	0.20	3/12	0.25	0.13	7/39	0.18	0.06	-0.07	n.s.
INNO1RAD	0.24	4/12	0.33	0.14	28/39	0.20	0.06	-0.13	n.s.
Innotype 2									
INNO2INC	0.17	2/14	0.14	0.10	7/38	0.18	0.06	0.04	n.s.
INNO2MOD	0.29	2/14	0.14	0.10	13/38	0.34	0.08	0.20	n.s.
INNO2ARC	0.40	7/14	0.50	0.14	14/38	0.37	0.08	-0.13	n.s.
INNO2RAD	0.44	8/14	0.57	0.14	15/38	0.39	0.08	-0.18	n.s.
Timing of market entry									
MARKT1ST	0.29	6/15	0.40	0.13	10/40	0.25	0.07	-0.15	n.s.
MARKT2ND	0.45	8/15	0.53	0.13	17/40	0.43	0.08	-0.11	n.s.
MARKT3RD	0.25	1/15	0.07	0.07	13/40	0.33	0.08	0.26	n.s.
Organizational independence									
ORGSEP	0.06	1/15	0.07	0.07	2/38	0.05	0.04	-0.01	n.s.
ORGDIV	0.07	1/15	0.07	0.07	3/39	0.08	0.04	0.01	n.s.
ORGINTRN	0.87	13/15	0.87	0.09	33/38	0.87	0.06	0.00	n.s.
Business model									
BMODNEW	0.08	1/15	0.07	0.07	3/35	0.09	0.05	0.02	n.s.
BMODSIM	0.28	1/15	0.07	0.07	13/35	0.37	0.08	0.31	n.s.
BMODOLD	0.64	13/15	0.87	0.09	19/35	0.54	0.09	-0.32	n.s.
Acquisitions									
TAKEOVER [K-S Z-test]	155.26	12[4]	203.33	166.27	26[4]	130.77	68.34	-72.56	n.s.
Alliances & Joint ventures									
ALLIANCE	0.27	5/15	0.33	0.13	8/34	0.24	0.07	-0.10	n.s.
Changing the value chain									
OUTOLD	3.10	15[5]	3.07	0.25	33[5]	3.12	0.19	0.05	n.s.
INNEW [K-S Z-test]	2.94	15[4]	2.73	0.30	33[4]	3.03	0.20	0.30	n.s.

[1] Average difference in mean values. [2] Significance levels of differences in mean values: *** <= 0.01, **<= 0.05, *<= 0.1, + <= 0.15, n.s. = not significant.
[3] Relative frequency. [4] Quantities n_1 and n_2 for interval-scaled parameters. [5] Quantities n_1 and n_2 for interval-scaled parameters for which normal distribution was not rejected.
[K-S Z-test] Variables tested by means of the Kolmogorov-Smirnov Z-test in case of no evidence of normal distribution (significance levels of t-tests in brackets).
All other variables tested by means of the t-test and checked with the Levene-Test for variance equality.

Table A.5.20'a': Strategic decisions - comparison of winners and losers of recent shock (wave 3, all product types)

Group and variable	'All' Mean	'Losers' R. freq.1[3]	Mean1	$s_1/\sqrt{n_1-1}$	'Winners' R. freq.2[3]	Mean2	$s_2/\sqrt{n_2-1}$	Difference of Means[1]	Significance level[2] p
Innotype 1									
INNO1INC	0.14	2/25	0.08	0.05	5/24	0.21	0.08	0.13	n.s.
INNO1MOD	0.53	18/25	0.72	0.09	8/24	0.33	0.09	-0.39*	0.051
INNO1ARC	0.16	3/25	0.12	0.07	5/24	0.21	0.08	0.09	n.s.
INNO1RAD	0.22	3/25	0.12	0.07	8/24	0.33	0.10	0.21	n.s.
Innotype 2									
INNO2INC	0.16	2/24	0.08	0.06	6/25	0.24	0.09	0.16	n.s.
INNO2MOD	0.33	4/24	0.17	0.08	12/25	0.48	0.10	0.31	n.s.
INNO2ARC	0.37	12/24	0.50	0.10	6/25	0.24	0.09	-0.26	n.s.
INNO2RAD	0.43	11/24	0.46	0.10	10/25	0.40	0.10	-0.06	n.s.
Timing of market entry									
MARKT1ST	0.31	8/24	0.33	0.10	8/28	0.29	0.09	-0.05	n.s.
MARKT2ND	0.42	14/24	0.58	0.10	8/28	0.29	0.09	-0.30	n.s.
MARKT3RD	0.27	2/24	0.08	0.06	12/28	0.43	0.10	0.35*	0.092
Organizational independence									
ORGSEP	0.06	1/24	0.04	0.04	2/26	0.07	0.05	0.04	n.s.
ORGDIV	0.08	2/24	0.08	0.06	2/27	0.07	0.05	-0.01	n.s.
ORGINTRN	0.88	22/24	0.92	0.06	22/26	0.85	0.07	-0.07	n.s.
Business model									
BMODNEW	0.11	2/21	0.10	0.07	3/26	0.12	0.06	0.02	n.s.
BMODSIM	0.32	4/21	0.19	0.09	11/26	0.42	0.10	0.23	n.s.
BMODOLD	0.57	15/21	0.71	0.10	12/26	0.49	0.10	-0.22	n.s.
Acquisitions									
TAKEOVER [K-S Z-test]	159.46	21[4]	185.71	115.73	16[4]	125.00	66.14	-60.71	n.s.
Alliances & Joint ventures									
ALLIANCE	0.24	5/22	0.23	0.09	6/24	0.25	0.09	0.02	n.s.
Changing the value chain									
OUTOLD	3.14	21[5]	3.14	0.21	23[5]	3.13	0.25	-0.01	n.s.
INNEW [K-S Z-test]	2.93	21[4]	2.52	0.21	23[4]	3.30	0.25	0.78	n.s.

[1] Average difference in mean values. [2] Significance levels of differences in mean values: *** <= 0.01, ** <= 0.05, *<= 0.1, + <= 0.15, n.s. = not significant.
[3] Relative frequency. [4] Quantities n_1 and n_2 for interval-scaled parameters. [5] Quantities n_1 and n_2 for interval-scaled parameters for which normal distribution was not rejected.
[K-S Z-test] Variables tested by means of the Kolmogorov-Smirnov Z-test in case of no evidence of normal distribution (significance levels of t-tests in brackets).
All other variables tested by means of the t-test and checked with the Levene-Test for variance equality.

Table A.5.21'p': Strategic decisions - comparison of winners and losers of all shocks (waves 1 - 3, without presses)

Group and variable	'All' Mean	'Losers' R. freq.1[3]	Mean$_1$	$s_1/\sqrt{n_1 - 1}$	'Winners' R. freq.2[3]	Mean$_2$	$s_2/\sqrt{n_2 - 1}$	Difference of Means[1]	Significance level[2] p
Innotype 1									
INNO1INC	0.14	2/30	0.07	0.05	5/21	0.24	0.10	0.17	n.s.
INNO1MOD	0.50	20/30	0.67	0.09	5/21	0.24	0.10	-0.43**	0.021
INNO1ARC	0.20	5/30	0.17	0.07	5/21	0.24	0.10	0.07	n.s.
INNO1RAD	0.25	5/30	0.17	0.07	8/21	0.38	0.11	0.21	n.s.
Innotype 2									
INNO2INC	0.17	4/31	0.13	0.06	5/21	0.24	0.10	0.11	n.s.
INNO2MOD	0.29	5/31	0.16	0.07	10/21	0.48	0.11	0.32	n.s.
INNO2ARC	0.38	17/31	0.55	0.09	3/21	0.14	0.07	-0.41**	0.033
INNO2RAD	0.44	14/31	0.45	0.09	9/21	0.43	0.11	-0.02	n.s.
Timing of market entry									
MARKT1ST	0.29	11/32	0.34	0.09	5/23	0.22	0.09	-0.13	n.s.
MARKT2ND	0.45	18/32	0.56	0.09	7/23	0.31	0.10	-0.26	n.s.
MARKT3RD	0.25	2/24	0.10	0.05	11/23	0.48	0.11	0.39**	0.038
Organizational independence									
ORGSEP	0.07	2/32	0.06	0.04	2/22	0.09	0.06	0.03	n.s.
ORGDIV	0.09	3/32	0.09	0.05	2/23	0.09	0.06	-0.01	n.s.
ORGINTRN	0.85	28/32	0.88	0.06	18/22	0.82	0.08	-0.06	n.s.
Business model									
BMODNEW	0.10	2/29	0.07	0.05	3/22	0.14	0.07	0.07	n.s.
BMODSIM	0.29	4/29	0.14	0.07	11/22	0.50	0.11	0.36*	0.075
BMODOLD	0.61	23/29	0.79	0.08	8/22	0.36	0.11	-0.43**	0.020
Acquisitions									
TAKEOVER K-S Z-test	147.50	26[4]	165.38	94.45	14[4]	114.29	72.52	-92.86	n.s.
Alliances & Joint ventures									
ALLIANCE	0.26	8/30	0.27	0.08	5/20	0.25	0.10	-0.02	n.s.
Changing the value chain									
OUTOLD K-S Z-test	3.08	29[4]	3.00	0.17	19[4]	3.21	0.29	0.21	n.s.
INNEW K-S Z-test	2.92	29[4]	2.72	0.19	19[4]	3.21	0.29	0.49	n.s.

[1] Average difference in mean values. [2] Significance levels of differences in mean values: *** = 0.01, ** = 0.05, * = 0.1, + = 0.15, n.s. = not significant.
[3] Relative frequency. [4] Quantities n_1 and n_2 for interval-scaled parameters.
K-S Z-test Variables tested by means of the Kolmogorov-Smirnov Z-test in case of no evidence of normal distribution (significance levels of t-tests in brackets).
All other variables tested by means of the t-test and checked with the Levene-Test for variance equality.

Table A.5.22'p': Strategic decisions - comparison of winners and losers of early shocks (waves 1 and 2, without presses)

Group and variable	'All'	'Losers'			'Winners'			Difference of Means[1]	Significance level[2] p
	Mean	R. freq.$_1$[3]	$Mean_1$	$s_1/\sqrt{n_1-1}$	R. freq.$_2$[3]	$Mean_2$	$s_2/\sqrt{n_2-1}$		
Innotype 1									
INNO1INC	0.14	0/12	0	0	7/37	0.19	0.07	0.19	n.s.
INNO1MOD	0.49	6/12	0.50	0.15	18/37	0.49	0.09	-0.01	n.s.
INNO1ARC	0.20	3/12	0.25	0.13	7/37	0.19	0.07	-0.06	n.s.
INNO1RAD	0.24	4/12	0.33	0.14	8/37	0.22	0.07	-0.12	n.s.
Innotype 2									
INNO2INC	0.16	2/14	0.14	0.10	6/36	0.17	0.06	0.02	n.s.
INNO2MOD	0.26	2/14	0.14	0.10	11/36	0.31	0.08	0.16	n.s.
INNO2ARC	0.40	7/14	0.50	0.14	13/36	0.36	0.08	-0.14	n.s.
INNO2RAD	0.46	8/14	0.57	0.14	15/36	0.42	0.08	-0.16	n.s.
Timing of market entry									
MARKT1ST	0.28	6/15	0.40	0.13	9/38	0.24	0.07	-0.16	n.s.
MARKT2ND	0.47	8/15	0.53	0.13	17/38	0.45	0.08	-0.09	n.s.
MARKT3RD	0.25	1/15	0.07	0.07	12/38	0.32	0.08	0.25	n.s.
Organizational independence									
ORGSEP	0.06	1/15	0.07	0.07	2/37	0.05	0.04	-0.01	n.s.
ORGDIV	0.07	1/15	0.07	0.07	3/38	0.08	0.04	0.01	n.s.
ORGINTRN	0.87	3/15	0.87	0.09	32/37	0.86	0.06	0.00	n.s.
Business model									
BMODNEW	0.08	1/15	0.07	0.07	3/34	0.09	0.05	0.02	n.s.
BMODSIM	0.29	1/15	0.07	0.07	13/34	0.38	0.08	0.32	n.s.
BMODOLD	0.63	13/15	0.87	0.09	18/34	0.53	0.09	-0.34	n.s.
Acquisitions									
TAKEOVER $^{K\text{-}S\ Z\text{-}test}$	155.26	12^4	208.33	166.27	26^4	130.77	68.34	-139.99	n.s.
Alliances & Joint ventures									
ALLIANCE	0.27	5/15	0.33	0.13	8/33	0.24	0.08	-0.09	n.s.
Changing the value chain									
OUTOLD	3.11	15^5	0.96	0.92	32^5	3.13	0.20	0.17	n.s.
INNEW $^{K\text{-}S\ Z\text{-}test}$	2.94	15^4	1.16	1.35	32^4	3.03	0.20	1.87	n.s.

[1] Average difference in mean values. [2] Significance levels of differences in mean values: *** <= 0.01, **<= 0.05, *< = 0.1, $^+$ <= 0.15, n.s. = not significant.
[3] Relative frequency. [4] Quantities n_1 and n_2 for interval-scaled parameters. [5] Quantities n_1 and n_2 for interval-scaled parameters for which normal distribution was not rejected.
$^{K\text{-}S\ Z\text{-}test}$ Variables tested by means of the Kolmogorov-Smirnov Z-test in case of no evidence of normal distribution (significance levels of t-tests in brackets).
All other variables tested by means of the t-test and checked with the Levene-Test for variance equality.

Table A.5.23'p': Strategic decisions - comparison of winners and losers of recent shock (wave 3, without presses)

Group and variable	'All' Mean	'Losers' R. freq.$_1$[3]	'Losers' Mean$_1$	'Losers' $s_1/\sqrt{n_1-1}$	'Winners' R. freq.$_2$[3]	'Winners' Mean$_2$	'Winners' $s_2/\sqrt{n_2-1}$	Difference of Means[1]	Significance level[2] p
Innotype 1									
INNO1INC	0.15	2/25	0.08	0.06	5/22	0.23	0.09	0.01	n.s.
INNO1MOD	0.51	18/25	0.72	0.09	6/22	0.27	0.10	-0.45**	0.019
INNO1ARC	0.17	3/25	0.12	0.06	5/22	0.23	0.09	0.11	n.s.
INNO1RAD	0.23	3/25	0.12	0.06	8/22	0.36	0.11	0.24	n.s.
Innotype 2									
INNO2INC	0.15	2/24	0.08	0.06	5/23	0.22	0.09	0.13	n.s.
INNO2MOD	0.30	4/24	0.17	0.08	10/23	0.43	0.11	0.27	n.s.
INNO2ARC	0.36	12/24	0.50	0.10	5/23	0.22	0.09	-0.28	n.s.
INNO2RAD	0.45	11/24	0.46	0.10	10/23	0.43	0.11	-0.02	n.s.
Timing of market entry									
MARKT1ST	0.30	8/24	0.33	0.10	7/26	0.27	0.09	-0.06	n.s.
MARKT2ND	0.44	14/24	0.58	0.10	8/26	0.31	0.09	-0.28	n.s.
MARKT3RD	0.26	2/24	0.08	0.06	11/26	0.42	0.10	0.34+	0.112
Organizational independence									
ORGSEP	0.06	1/24	0.04	0.04	2/25	0.08	0.06	0.04	n.s.
ORGDIV	0.08	2/24	0.08	0.06	2/26	0.08	0.05	-0.01	n.s.
ORGINTRN	0.88	22/24	0.92	0.06	21/25	0.84	0.07	-0.08	n.s.
Business model									
BMODNEW	0.11	2.21	0.10	0.07	3/25	0.12	0.07	0.03	n.s.
BMODSIM	0.33	4/21	0.19	0.09	11/25	0.44	0.10	0.25	n.s.
BMODOLD	0.57	15/21	0.71	0.10	11/25	0.44	0.10	-0.27	n.s.
Acquisitions									
TAKEOVER K-S Z-test	159.46	21[4]	185.71	115.73	16[4]	125.00	66.14	-60.71	n.s.
Alliances & Joint ventures									
ALLIANCE	0.24	5/22	0.23	0.09	6/23	0.26	0.09	0.03	n.s.
Changing the value chain									
OUTOLD	3.14	21[5]	3.14	0.21	22[5]	3.14	0.26	0.00	n.s.
INNEW K-S Z-test	2.93	21[4]	2.52	0.21	22[4]	3.32	0.27	0.80	n.s.

[1] Average difference in mean values. [2] Significance levels of differences in mean values: *** <= 0.01, ** <= 0.05, * <= 0.1, + <= 0.15, n.s. = not significant.
[3] Relative frequency. [4] Quantities n_1 and n_2 for interval-scaled parameters. [5] Quantities n_1 and n_2 for interval-scaled parameters for which normal distribution was not rejected.
K-S Z-test Variables tested by means of the Kolmogorov-Smirnov Z-test in case of no evidence of normal distribution (significance levels of t-tests in brackets).
All other variables tested by means of the t-test and checked with the Levene-Test for variance equality.

Table A.5.24'a': Operational decisions - comparison of winners and losers of all shocks (waves 1 - 3, all product types)

Group and variable	'All'	'Losers'			'Winners'			Difference of Means[1]	Significance level[2] p
	Mean	n_1	$Mean_1$	$s_1/\sqrt{n_1-1}$	n_2	$Mean_2$	$s_2/\sqrt{n_2-1}$		
Resource commitment									
ENGTR70	0.07	23	0.06	0.02	14	0.08	0.05	0.02	n.s.
ENGTR00	0.04	21	0.05	0.08	19	0.31	0.07	0.17+	0.101
ENGBR70	0.008	21	0.006	0.003	11	0.01	0.09	0.007	n.s
ENGBR00	0.09	20	0.10	0.03	17	0.08	0.02	-0.02	n.s
ORGBEGIN	21.07	18	17.00	5.71	11	27.73	8.83	10.73	n.s
ORGEND	77.73	20	78.65	6.56	10	75.90	11.63	-2.75	n.s
ORGGROW	13.44	17	19.03	7.73	10	3.92	0.73	-15.11*	0.069
Level of hierarchy									
HIERLEV	2.99	27	2.98	0.20	14	3.00	0.26	0.02	n.s.
HIERBROA	7.31	20	7.50	1.24	9	6.89	1.72	-0.61	n.s
Extent of decisions									
DECISION	3.96	28	4.07	0.18	20	3.80	0.22	-0.27	n.s.
Extent of implementation program									
REALISAT	3.60	31	3.55	0.20	22	3.68	0.25	0.11	n.s.
Cooperation with suppliers and customers									
SUPCOOP	3.40	32	3.34	0.23	23	3.48	0.21	0.14	n.s.
SUPCONTR	3.64	33	3.61	0.19	23	3.70	0.22	0.09	n.s
CLICOOP	3.53	32	3.59	0.16	23	3.43	0.23	-0.16	n.s
CLICONTR	3.51	32	3.69	0.23	23	3.26	0.27	-0.43	n.s
Trainings									
TCPRC2	0.07	18	0.07	0.03	13	0.08	0.05	0.01	n.s
TCPRC4	0.07	20	0.08	0.03	14	0.07	0.02	-0.01	n.s
TCPRC6	0.02	20	0.02	0.02	13	0.03	0.02	-0.009	n.s
BSPRC2	0.04	21	0.03	0.01	13	0.06	0.05	0.03	n.s
BSPRC4	0.05	21	0.04	0.04	11	0.05	0.03	0.01	n.s
BSPRC6	0.00	20	0.00	0.00	13	0.00	0.00	0.00	n.s
External R&D									
RDPURCHA	2.33	21	2.58	1.48	13	1.93	1.22	0.65	n.s
RDPURCR	0.03	20	0.04	0.02	13	0.02	0.007	-0.02	n.s
RDSAVE	16.97	17	9.47	3.63	12	27.58	20.40	-18.11	n.s
RDSAVR	0.06	16	0.03	0.02	12	0.09	0.07	0.06	n.s
New hires									
TCHIRE	12.12	21	14.71	3.35	12	7.58	2.72	-7.13(+)	0.155
BSHIRE	1.79	22	1.95	0.73	12	1.50	0.66	-0.45	n.s

[1] Average difference in mean values. [2] Significance levels of differences in mean values: *** <= 0.01, ** <= 0.05, * <= 0.1, + <= 0.15, n.s. = not significant.
K-S Z-test Variables tested by means of the Kolmogorov-Smirnov Z-test in case of no evidence of normal distribution (significance levels of t-tests in brackets).
All other variables tested by means of the t-test and checked with the Levene-Test for variance equality.

Table A.5.25'a': Operational decisions - comparison of winners and losers of early shocks (waves 1 and 2, all product types)

Group and variable	'All' Mean	'Losers' n_1	$Mean_1$	$s_1/\sqrt{n_1-1}$	'Winners' n_2	$Mean_2$	$s_2/\sqrt{n_2-1}$	Difference of Means[1]	Significance level[2] p
Resource commitment									
ENGTR70	0.07	12	0.05	0.02	25	0.08	0.03	0.03	n.s.
ENGTR00	0.41	10	0.06	0.12	28	0.04	0.06	-0.02	n.s.
ENGBR70	0.008	11	0.01	0.005	21	0.008	0.01	0.00	n.s
ENGBR00	0.09	9	0.11	0.04	26	0.09	0.02	-0.03	n.s
ORGBEGIN	21.11	12	11.25	4.03	16	28.50	7.94	17.25*	0.066
ORGEND	77.67	12	70.67	10.20	17	82.59	7.14	11.92	n.s
ORGGROW	13.80	11	17.97	8.88	15	10.74	6.41	-7.23	n.s
Level of hierarchy									
HIERLEV	3.01	14	2.82	0.29	25	3.12	0.20	0.30	n.s.
HIERBROA	7.59	10	4.70	0.96	17	9.29	1.42	4.59**	0.013
Extent of decisions									
DECISION	4.00	13	4.23	0.17	33	3.91	0.19	-0.30	n.s.
Extent of implementation program									
REALISAT	3.59	14	3.36	0.34	37	3.68	0.18	0.32	n.s.
Cooperation with suppliers and customers									
SUPCOOP	3.40	15	3.73	0.30	38	3.26	0.19	-0.47	n.s.
SUPCONTR	3.67	15	3.47	0.26	39	3.74	0.18	0.28	n.s
CLICOOP	3.49	15	3.60	0.24	38	3.45	0.17	-0.15	n.s
CLICONTR	3.53	15	3.47	0.35	38	3.55	0.21	0.08	n.s
Trainings									
TCPRC2	0.08	11	0.04	0.02	19	0.10	0.04	0.06	n.s.
TCPRC4	0.07	11	0.07	0.03	21	0.08	0.03	0.01	n.s
TCPRC6	0.02	11	0.00	0.00	20	0.03	0.02	0.03	n.s
BSPRC2	0.04	11	0.01	0.00	21	0.06	0.03	0.06	n.s
BSPRC4	0.05	11	0.01	0.01	19	0.07	0.05	0.06	n.s
BSPRC6	0.00	11	0.00	0.00	20	0.00	0.00	0.00	n.s
External R&D									
RDPURCHA	2.47	11	1.57	0.75	21	2.95	1.59	1.38	n.s
RDPURCR	0.03	11	0.04	0.02	21	0.03	0.02	-0.02	n.s
RDSAVE	17.22	9	8.44	5.63	18	21.61	13.67	13.17	n.s.
RDSAVR	0.03	9	0.08	0.03	18	0.04	0.01	-0.05	n.s.
New hires									
TCHIRE	12.74	11	20.45	5.55	20	8.50	1.96	-7.13(+)	0.155
BSHIRE	1.88	12	2.42	1.07	20	1.55	0.61	-0.97	n.s

[1] Average difference in mean values. [2] Significance levels of differences in mean values: *** <= 0.01, ** <= 0.05, *< = 0.1, + <= 0.15, n.s. = not significant.
K-S Z-test Variables tested by means of the Kolmogorov-Smirnov Z-test in case of no evidence of normal distribution (significance levels of t-tests in brackets).
All other variables tested by means of the t-test and checked with the Levene-Test for variance equality.

Table A.5.26'a': Operational decisions - comparison of winners and losers of recent shock (wave 3, all product types)

Group and variable	'All'	'Losers'			'Winners'			Difference of Means[1]	Significance level[2] p
	Mean	n_1	$Mean_1$	$s_1/\sqrt{n_1-1}$	n_2	$Mean_2$	$s_2/\sqrt{n_2-1}$		
Resource commitment									
ENGTR70	0.07	18	0.06	0.03	16	0.08	0.04	0.02	n.s.
ENGTR00	0.42	17	0.51	0.08	21	0.36	0.07	-0.15	n.s.
ENGBR70	0.008	17	0.007	0.003	12	0.01	0.008	0.005	n.s
ENGBR00	0.10	16	0.10	0.03	19	0.10	0.02	0.0009	n.s
ORGBEGIN	24.67	12	19.75	7.71	12	29.58	8.27	9.83	n.s
ORGEND	80.96	14	83.21	5.18	11	78.09	10.75	-5.12	n.s
ORGGROW	10.79	12	17.24	8.14	11	3.74	0.68	-13.45+	0.129
Level of hierarchy									
HIERLEV	3.03	20	3.00	0.25	16	3.06	0.23	0.06	n.s.
HIERBROA	7.72	14	8.21	1.65	11	7.09	1.42	-1.12	n.s
Extent of decisions									
DECISION	3.95	22	4.00	0.22	22	3.91	0.22	-0.09	n.s.
Extent of implementation program									
REALISAT	3.67	23	3.70	0.20	25	3.64	0.24	-0.06	n.s.
Cooperation with suppliers and customers									
SUPCOOP	3.38	24	3.25	0.27	26	3.50	0.19	0.25	n.s.
SUPCONTR	3.65	25	3.60	0.24	26	3.69	0.21	0.09	n.s
CLICOOP	3.56	24	3.67	0.17	26	3.46	0.22	-0.21	n.s
CLICONTR	3.54	24	3.71	0.26	26	3.38	0.25	-0.32	n.s
Trainings									
TCPRC2	0.07	13	0.08	0.03	15	0.07	0.04	-0.008	n.s
TCPRC4	0.08	15	0.08	0.04	16	0.08	0.02	-0.004	n.s
TCPRC6	0.03	15	0.03	0.03	15	0.03	0.01	-0.002	n.s
BSPRC2	0.04	16	0.03	0.02	15	0.05	0.04	0.02	n.s
BSPRC4	0.05	16	0.05	0.05	13	0.05	0.03	-0.003	n.s
BSPRC6	0.00	15	0.00	0.00	15	0.00	0.00	0.00	n.s
External R&D									
RDPURCHA	2.47	17	2.77	1.80	15	2.13	1.09	-0.64	n.s
RDPURCR	0.04	16	0.04	0.02	15	0.03	0.01	-0.006	n.s
RDSAVE	17.57	14	7.64	3.12	14	27.50	17.54	19.86	n.s
RDSAVR	0.06	13	0.02	0.009	14	0.10	0.06	0.08	n.s
New hires									
TCHIRE	11.93	16	14.94	4.29	14	8.50	2.48	-6.44	n.s
BSHIRE	1.48	17	1.53	0.80	14	1.43	0.57	-0.10	n.s

[1] Average difference in mean values. [2] Significance levels of differences in mean values: *** <= 0.01, ** <= 0.05, * < = 0.1, + <= 0.15, n.s. = not significant.
K-S Z-test Variables tested by means of the Kolmogorov-Smirnov Z-test in case of no evidence of normal distribution (significance levels of t-tests in brackets).
All other variables tested by means of the t-test and checked with the Levene-Test for variance equality.

Table A.5.27'p': Operational decisions - comparison of winners and losers of all shocks (waves 1 - 3, without presses)

Group and variable	'All'	'Losers'			'Winners'			Difference of Means[1]	Significance level[2] p
	Mean	n_1	$Mean_1$	$s_1/\sqrt{n_1-1}$	n_2	$Mean_2$	$s_2/\sqrt{n_2-1}$		
Resource commitment									
ENGTR70	0.07	23	0.06	0.02	14	0.08	0.05	0.03	n.s.
ENGTR00	0.40	21	0.05	0.08	19	0.31	0.07	0.17+	0.101
ENGBR70	0.008	21	0.006	0.003	11	0.01	0.09	0.007	n.s
ENGBR00	0.09	20	0.10	0.03	17	0.08	0.02	-0.02	n.s
ORGBEGIN	21.07	18	17.00	5.71	11	27.73	8.83	10.73	n.s
ORGEND	77.73	20	78.65	6.56	10	75.90	11.63	-2.75	n.s
ORGGROW	13.44	17	19.03	7.73	10	3.92	0.73	-15.11*	0.069
Level of hierarchy									
HIERLEV	2.94	27	2.98	0.20	14	3.00	0.26	0.02	n.s.
HIERBROA	7.39	20	7.50	1.24	9	6.89	1.72	-0.38	n.s
Extent of decisions									
DECISION	3.96	28	4.07	0.18	20	3.80	0.22	-0.27	n.s.
Extent of implementation program									
REALISAT	3.60	31	3.55	0.20	22	3.68	0.25	0.13	n.s.
Cooperation with suppliers and customers									
SUPCOOP	3.41	22	3.50	0.22	32	3.34	0.23	-0.16	n.s.
SUPCONTR	3.65	22	3.73	0.23	33	3.61	0.19	-0.12	n.s
CLICOOP	3.54	22	3.45	0.24	32	3.59	0.16	0.14	n.s
CLICONTR	3.52	22	3.27	0.28	32	3.69	0.23	0.41	n.s
Trainings									
TCPRC2	0.08	18	0.07	0.03	13	0.08	0.05	0.01	n.s
TCPRC4	0.08	20	0.08	0.03	14	0.07	0.02	-0.007	n.s
TCPRC6	0.02	20	0.02	0.02	13	0.03	0.02	0.01	n.s
BSPRC2	0.04	21	0.03	0.01	13	0.06	0.05	0.03	n.s
BSPRC4	0.05	21	0.04	0.04	11	0.05	0.03	0.01	n.s
BSPRC6	0.00	20	0.00	0.00	13	0.00	0.00	0.00	n.s
External R&D									
RDPURCHA	2.33	21	2.58	1.48	13	1.93	1.22	0.65	n.s
RDPURCR	0.03	20	0.04	0.02	13	0.02	0.007	-0.02	n.s
RDSAVE	16.97	17	9.47	3.63	12	27.58	20.40	-18.11	n.s
RDSAVR	0.06	16	0.03	0.02	12	0.09	0.07	0.06	n.s
New hires									
TCHIRE	12.12	21	14.71	3.35	12	7.58	2.72	-7.13(+)	0.155
BSHIRE	1.79	22	1.95	0.73	12	1.50	0.66	-0.45	n.s

[1] Average difference in mean values. [2] Significance levels of differences in mean values: *** <= 0.01, **<= 0.05, *< = 0.1, + <= 0.15, n.s. = not significant.
K-S Z-test Variables tested by means of the Kolmogorov-Smirnov Z-test in case of no evidence of normal distribution (significance levels of t-tests in brackets).
All other variables tested by means of the t-test and checked with the Levene-Test for variance equality.

Table A.5.28'p': Operational decisions - comparison of winners and losers of early shocks (waves 1 and 2, without presses)

Group and variable	'All'	'Losers'			'Winners'			Difference of Means[1]	Significance level[2] p
	Mean	n_1	$Mean_1$	$s_1/\sqrt{n_1-1}$	n_2	$Mean_2$	$s_2/\sqrt{n_2-1}$		
Resource commitment									
ENGTR70	0.07	12	0.05	0.02	24	0.08	0.03	0.03	n.s.
ENGTR00	0.41	10	0.06	0.12	28	0.04	0.06	-0.02	n.s.
ENGBR70	0.008	11	0.01	0.005	21	0.008	0.005	0.003	n.s
ENGBR00	0.09	9	0.11	0.04	26	0.09	0.02	-0.03	n.s
ORGBEGIN	21.11	12	11.25	4.03	16	28.50	7.94	17.25*	0.066
ORGEND	77.66	12	70.67	10.20	17	82.59	7.14	11.92	n.s
ORGGROW	13.80	11	17.97	8.88	15	10.74	6.41	-7.23	n.s
Level of hierarchy									
HIERLEV	2.96	14	2.82	0.29	24	3.04	0.29	-0.22	n.s.
HIERBROA	7.69	10	4.70	0.96	16	9.56	1.49	4.86**	0.011
Extent of decisions									
DECISION	4.00	13	4.23	0.17	32	3.91	0.20	-0.32	n.s.
Extent of implementation program									
REALISAT	3.58	14	3.36	0.34	36	3.67	0.19	0.31	n.s.
Cooperation with suppliers and customers									
SUPCOOP	3.40	15	3.73	0.30	37	3.27	0.20	-0.46	n.s.
SUPCONTR	3.68	15	3.47	0.26	38	3.76	0.18	0.30	n.s
CLICOOP	3.50	15	3.60	0.24	37	3.46	0.17	-0.14	n.s
CLICONTR	3.54	15	3.47	0.35	37	3.57	0.22	0.10	n.s
Trainings									
TCPRC2	0.08	11	0.04	0.02	18	0.10	0.04	0.06	n.s.
TCPRC4	0.07	11	0.07	0.03	2	0.08	0.03	0.01	n.s
TCPRC6	0.02	11	0.0006	0.0005	29	0.03	0.02	-0.03	n.s
BSPRC2	0.04	11	0.005	0.002	20	0.06	0.03	0.06+	0.106
BSPRC4	0.05	11	0.008	0.005	18	0.07	0.05	0.06	n.s
BSPRC6	0.00	11	0.00	0.00	19	0.00	0.00	0.00	n.s
External R&D									
RDPURCHA	2.47	11	1.57	0.75	21	2.95	1.59	1.38	n.s
RDPURCR	0.03	11	0.04	0.02	21	0.03	0.02	-0.009	n.s
RDSAVE	17.22	9	8.44	5.63	18	21.61	13.67	13.17	n.s.
RDSAVR	0.03	9	0.03	0.03	18	0.02	0.09	-0.01	n.s.
New hires									
TCHIRE	12.74	11	20.45	5.55	20	8.50	1.96	-11.95*	0.064
BSHIRE	1.88	12	2.42	1.07	20	1.55	0.61	-0.87	n.s

[1] Average difference in mean values. [2] Significance levels of differences in mean values: ***<= 0.01, < 0.05, * <= 0.1, + <= 0.15, n.s. = not significant.

K-S Z-test Variables tested by means of the Kolmogorov-Smirnov Z-test in case of no evidence of normal distribution (significance levels of t-tests in brackets).

All other variables tested by means of the t-test and checked with the Levene-Test for variance equality.

Table A.5.29'p': Operational decisions - comparison of winners and losers of recent shock (wave 3, without presses)

Group and variable	'All'	'Losers'			'Winners'			Difference of Means[1]	Significance level[2] p
	Mean	n_1	$Mean_1$	$s_1/\sqrt{n_1-1}$	n_2	$Mean_2$	$s_2/\sqrt{n_2-1}$		
Resource commitment									
ENGTR70	0.07	18	0.06	0.03	15	0.08	0.04	0.02	n.s.
ENGTR00	0.42	17	0.51	0.08	21	0.36	0.07	-0.15	n.s.
ENGBR70	0.009	17	0.01	0.003	12	0.01	0.008	0.00	n.s
ENGBR00	0.10	16	0.10	0.03	19	0.10	0.02	0.00	n.s
ORGBEGIN	24.67	12	19.75	7.71	12	29.58	8.27	9.83	n.s
ORGEND	80.96	14	83.21	5.18	11	78.09	10.75	-5.12	n.s
ORGGROW	10.79	12	17.24	8.14	11	3.74	0.68	-13.50+	0.126
Level of hierarchy									
HIERLEV	2.97	20	3.00	0.25	15	2.93	0.21	0.06	n.s.
HIERBROA	7.83	14	8.21	1.65	10	7.30	1.56	-0.91	n.s
Extent of decisions									
DECISION	3.95	22	4.00	0.22	21	3.90	0.23	-0.09	n.s.
Extent of implementation program									
REALISAT	3.66	23	3.70	0.20	24	3.63	0.25	-0.06	n.s.
Cooperation with suppliers and customers									
SUPCOOP	3.39	24	3.25	0.27	25	3.52	0.19	0.27	n.s.
SUPCONTR	3.66	25	3.60	0.24	25	3.72	0.22	0.12	n.s
CLICOOP	3.57	24	3.67	0.17	25	3.48	0.22	-0.19	n.s
CLICONTR	3.55	24	3.71	0.26	25	3.40	0.26	-0.31	n.s
Trainings									
TCPERC2	0.07	13	0.08	0.03	14	0.08	0.05	0.00	n.s
TCPERC4	0.08	15	0.08	0.04	15	0.08	0.02	-0.01	n.s
TCPERC6	0.03	15	0.03	0.03	14	0.03	0.02	0.00	n.s
BSPERC2	0.04	16	0.03	0.02	14	0.05	0.04	0.02	n.s
BSPERC4	0.05	16	0.05	0.05	12	0.05	0.03	0.00	n.s
BSPERC6	0.00	15	0.00	0.00	14	0.00	0.00	0.00	n.s
External R&D									
RDPURCHA	2.47	17	2.77	1.80	15	2.13	1.09	-0.64	n.s
RDPURCR	0.04	16	0.04	0.02	15	0.03	0.01	-0.01	n.s
RDSAVE	17.57	14	7.64	3.12	14	27.50	17.54	19.86	n.s
RDSAVR	0.06	13	0.02	0.009	14	0.10	0.06	0.08	n.s
New hires									
TCHIRE	11.93	16	14.94	4.29	14	8.50	2.48	-6.44	n.s
BSHIRE	1.48	17	1.53	0.80	14	1.43	0.57	-0.10	n.s

[1] Average difference in mean values. [2] Significance levels of differences in mean values: ***<= 0.01, < 0.05, * <= 0.1, + <= 0.15, n.s. = not significant.
[K-S Z-test] Variables tested by means of the Kolmogorov-Smirnov Z-test in case of no evidence of normal distribution (significance levels of t-tests in brackets).
All other variables tested by means of the t-test and checked with the Levene-Test for variance equality.

Appendix of chapter 6

Table A.6.1: Spearman correlation of company attributes

#	Group and construct	Correlation with # 2	3	4	5	6	7	8	9	10	11
1	Ability to forecast techno. FOREC	-0.38***	0.10	0.21+	0.44***	0.17	0.45***	0.18	-0.10	0.27**	-0.22+
2	Resistance to innovation RESISTAN		-0.19	0.05	-0.10	-0.23+	-0.36***	-0.34**	0.24	-0.23*	-0.04
3	Flexibility of plan FLEX			0.14	0.19	0.04	0.42***	0.46***	-0.40**	-0.008	0.12
4	Information flow INF				-0.01	-0.04	0.09	0.17	-0.39**	0.21	0.03
5	Company culture CULT					-0.02	0.47***	0.38***	-0.06	0.06	0.10
6	Distribution of power POWER						0.01	-0.06	-0.14	0.12	-0.27*
7	Management focus FOCUS							0.45***	-0.03	0.08	0.20
8	Consequence of decision CONSEQUE								-0.08	0.28**	0.30**
9	Financial base FINANCE									0.17	0.24
10	In-house manufacturing MANUF										0.19
11	Attractivity for new hires ATTRACT										

Significance levels of Spearman correlation coefficients:
*** <= 0.01, ** <= 0.05, * <= 0.10, + <= 0.15.

Table A.6.2: Spearman correlation of strategic decision variables

#	Group and construct	Correlation with #							
		2	3	4	5	6	7	8	9
1	Degree of innovation 1 INNO1	-0.17	0.09	0.18	0.26*	-0.14	0.04	-0.11	-0.11
2	Degree of innovation 2 INNO2		-0.20	-0.29	-0.04	0.06	-0.02	0.14	0.07
3	Timing of market entry MARKT			0.21+	0.23+	-0.32**	-0.15	-011	-0.04
4	Organizational independence ORG				0.32**	0.03	0.26*	-0.22+	0.09
5	Business model BMOD					-0.01	0.12	0.12	0.25*
6	Acquisitions TAKEOVER						0.36**	0.11	-0.02
7	Alliances & Joint ventures ALLIANCE							-0.05	0.09
8	Change in value chain OUTOLD								0.22+
9	INNEW								

Significance levels of Spearman correlation coefficients:
*** <= 0.01, ** <= 0.05, * <= 0.10, + <= 0.15.

Table A.6.3: Spearman correlation of operational decision variables

#	Group and construct	Correlation with # 2	3	4	5	6	7	8	9	10	11	12
	Resource commitment											
1	ENG70	-0.12	0.04	0.08	0.04	-0.006	0.02	0.14	0.30+	0.19	-0.28+	0.39**
2	ENG00		-0.28	0.23	0.08	0.05	-0.46***	0.09	0.09	0.26	0.07	-0.23
3	ORGPERC			0.34*	0.22	-0.13	0.34**	0.23	-0.15	0.38*	-0.19	0.21
	Level of hierarchy											
4	HIER				-0.02	0.03	-0.05	0.18	0.08	0.27+	-0.40**	0.001
	Extent of decisions											
5	DECISION					-0.04	-0.04	0.14	-0.14	0.00	0.29	-0.14
	Extent of implementation											
6	REALISAT						-0.07	0.02	-0.06	-0.14	-0.06	-0.06
	Cooperation suppl./ clients											
7	CLICOOP							0.04	-0.23	0.09	0.06	0.23
8	CONTROL								0.17	0.19	0.09	-0.06
	Trainings											
9	TRAINING									0.11	0.03	0.14
	External R&D											
10	RDEXTERN										0.12	-0.02
	New hires											
11	TCHIRE											0.16
12	BSHIRE											

Significance levels of Spearman correlation coefficients:
*** <= 0.01, ** <= 0.05, * <= 0.10, + <= 0.15.

Table 6.4: Kaplan-Meier graphs for categorical strategic variables

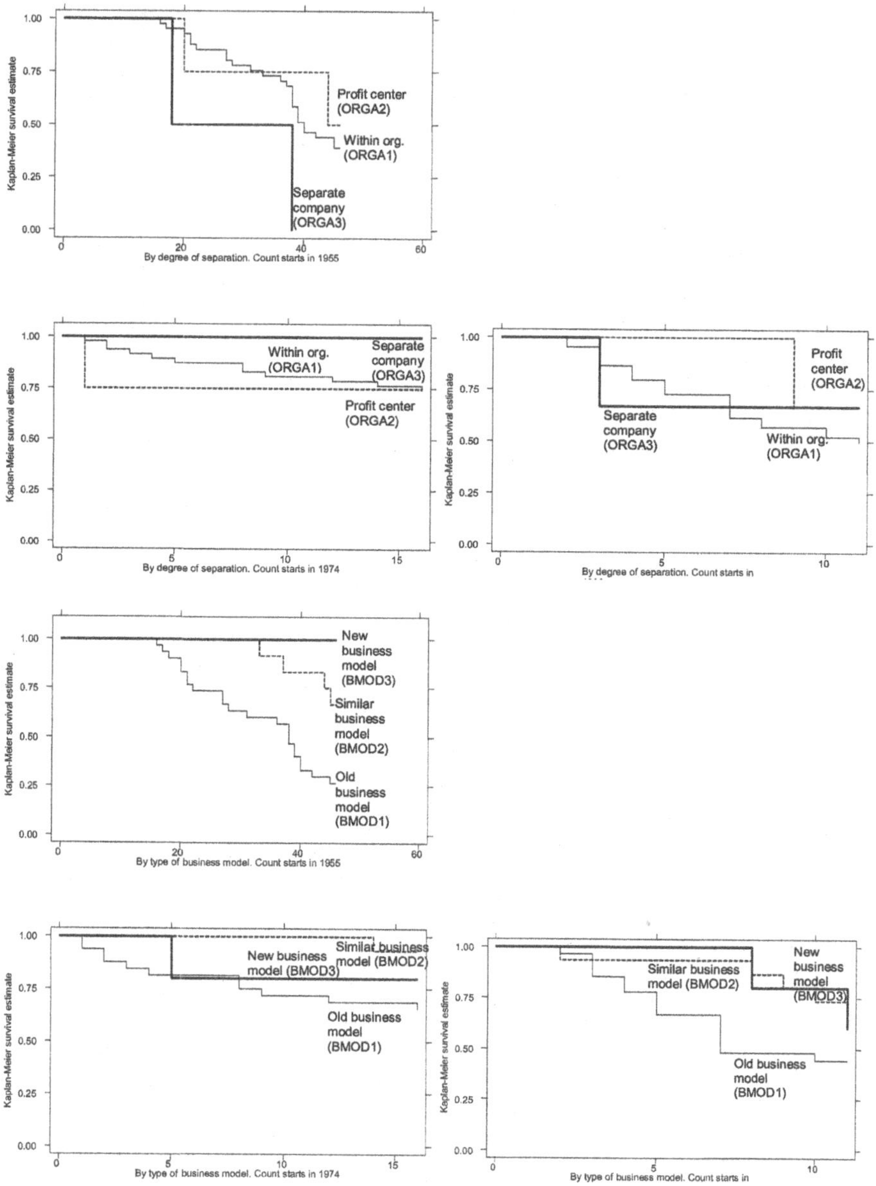

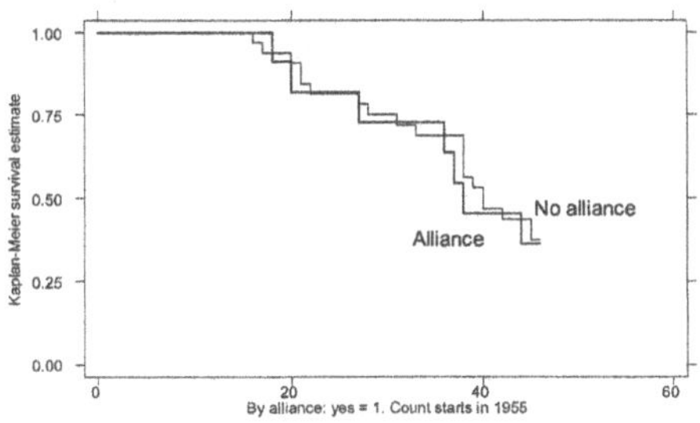

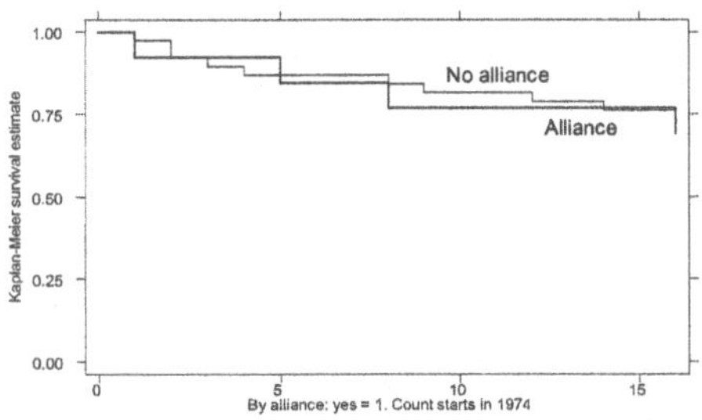

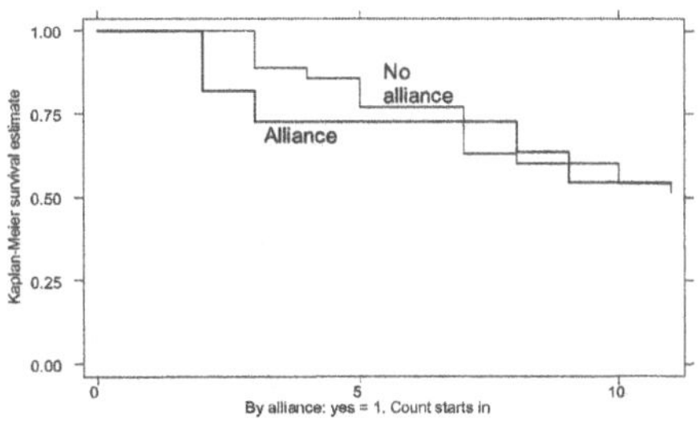

Table A.6.5: Significance of attributes once included in the basic model 1955 - 2001

Variable	Increment to R-squared	Significance (p-value)
FOREC	.0108	n.s.
RESISTAN	.0272	n.s.
FLEX	.0127	n.s.
INF	.0246	n.s.
CULT	.0306	n.s.
POWER	.0191	n.s.
IMPFINANCE	.0056	n.s.
MANUF	.0224	n.s.
IMPATTRACT	.0005	n.s.

Significance levels of differences of mean values:
*** p<=0.01, **p<=0.05, * p<=0.1, $^{+}$p<=0.15.
Note: OLS Regression model, standard error in brackets

Table A.6.6: Significance of attributes once included in the basic model 1974 - 1989

Variable	Increment to CHI-squared	Significance (p-value)
FOREC	3.95	$p < 0.05$
RESISTAN	1.77	n.s.
FLEX	1.52	n.s.
INF	0.84	n.s.
CULT	0.61	n.s.
POWER	1.18	n.s.
IMPFOCUS	1.50	n.s.
IMPFINANCE	0.27	n.s.
MANUF	1.16	n.s.
IMPATTRACT	3.70	$p < 0.1$

Significance levels of differences of mean values:
*** p<=0.01, **p<=0.05, * p<=0.1, +p<=0.15.
Note: MLE Cox model, standard error in brackets

Table A.6.7: Significance of attributes once included in the basic model 1990 - 2001

Variable	Increment to CHI-squared	Significance (p-value)
FOREC	1.17	n.s.
RESISTAN	0	n.s.
FLEX	-0.51	n.s.
INF	0.05	n.s.
CULT	2.66	$p < 0.15$
IMPFOCUS	3.25	$p < 0.1$
IMPCONSEQUE	0.21	n.s.
MANUF	0.09	n.s.
IMPATTRACT	0	n.s.

Significance levels of differences of mean values:
*** p<=0.01, **p<=0.05, * p<=0.1, +p<=0.15.
Note: MLE Cox model, standard error in brackets

Table A.6.8: Significance of strategic variables once included in the basic model 1955 - 2001

Variable	Increment to R-squared	Significance (p-value)
INNO1INC	.0286	n.s.
IINO1MOD	.0236	n.s.
INNO1ARC	.0135	n.s.
INNO1RAD	.0092	n.s.
INNO2RAD	.0206	n.s.
INNO2ARC	.0026	n.s.
INNO2INC	.0154	n.s.
MARKT1ST	.0124	n.s.
MARKT2ND	.0124	n.s.
ORGDIV	-.0031	n.s.
ORGINTRN	-.0171	n.s.
ORGSEP	-.0163	n.s.
BMODSIM	-.0295	$p < 0.1$

Table A.6.8 (continuation)

Variable	Increment to R-squared	Significance (p-value)
BMODNEW	-.0303	n.s.
IMPTAKEOVER	.0027	n.s.
ALLIANCE	-.0312	n.s.
OUTOLD	-.0408	n.s.
INNEW	-.0182	n.s.

Significance levels of differences of mean values:
*** p<=0.01, **p<=0.05, * p<=0.1, +p<=0.15.
Note: OLS Regression model, standard error in brackets

Table A.6.9: Significance of strategic variables once included in the basic model 1974 - 1989

Variable	Increment to CHI-squared	Significance (p-value)
MARKT3RD	n.a.	$p < 0.15$
BMODSIM	n.a.	$p < 0.1$
BMODOLD	n.a.	$p < 0.1$

Significance levels of differences of mean values:
*** p<=0.01, **p<=0.05, * p<=0.1, +p<=0.15.
Note: MLE Cox model, standard errors in brackets

Table A.6.9: Significance of strategic variables once included in the basic model 1990 - 2001

Variable	Increment to CHI-squared	Significance (p-value)
INNO1INC	1.28	n.s.
INNO1MOD	3.50	$p < 0.15$
INNO1ARC	0.63	n.s.
INNO1RAD	2.63	n.s.
INNO2INC	0.24	n.s.
INNO2ARC	1.02	n.s.
INNO2RAD	0.71	n.s.
MARKT2ND	0	n.s.
ORGDIV	1.87	n.s.
ORGINTRN	0.91	n.s.
ORGSEP	1.89	n.s.
BMODSIM	4.04	$p < 0.05$
BMODOLD	4.38	$p < 0.05$
BMODNEW	-0.74	n.s.
IMPTAKEOVER	0	n.s.
ALLIANCE	-2.87	n.s.
INNEW	-1.48	n.s.
OUTOLD	-0.23	n.s.

Significance levels of differences of mean values:
*** p<=0.01, **p<=0.05, * p<=0.1, +p<=0.15.
Note: MLE Cox model, standard errors in brackets

List of figures

List of tables

Abbreviations

AMT	American Machine Tool Association
ATC	automatic tool changing
CAD	computer aided design
CAM	computer aided manufacturing
CIM	computer integrated manufacturing
CNC	computer numerical control
GUI	graphical user interface
LSI	large scale integration
VLSI	very large scale integration
MIT	Massachusetts Institute of Technology
MITI	Japanese Ministry for International Trade and Industry
MLE	maximum likelihood estimator
NC	numerical control
OLS	ordinary least squares
PC	personal computer
PLC	product life cycle
PPS	production planning system
RBC	real business cycle
RMSE	root mean square error
US	United States of America
USDoC	United States Department of Commerce
VDW	Verein Deutscher Werkzeugmaschinenfabriken e.V.

References

Abernathy W, Clark K. 1985. Innovation: Mapping the Winds of Creative Destruction. *Research Policy* **14**: 3-22.

Abernathy WM, Utterback JM. 1978. Patterns of Industrial Innovation. *Technology Review* **80**(7): 41-47.

Ackermann C, Harrop J. 1985. The Management of Technological Innovation in the Machine Tool Industry: a Cross-national regional survey of Britain and Switzerland. *R&D Management* **15**(3): 207 - 218.

Afuah AN, Bahram N. 1995. The Hypercube of Innovation. *Research Policy* **24**: 51-76.

Albers S, Eggers S. 1991. Organisatorische Gestaltung von Produktinnovations-Prozessen - Führt der Wechsel des Organisationsgrades zu Innovationserfolg? *ZFBF* **43**(1): 44-64.

Allen TJ. 1984. *Managing the Flow of Innovation*: Cambridge.

Amburgey TL, Kelly D, Barnett W. 1993. Resetting the Clock: The Dynamics of Organizational Change and Failure. *Administrative Science Quarterly*(38): 51-73.

AMT. 1990. *The Economic Handbook of the Machine Tool Industry*. AMT - The Association for Manufacturing Technology: McLean, Virginia.

Anderson P, Tushman ML. 1990. Technological Discontinuities and Dominant Designs: A Cyclical Model of Technological Change. *Administrative Science Quarterly* **35**(4): 604-633.

Anonymous. 1980. Bring Technology Out of the Lab and Into Your Plant. *Production* **86**(2): 83.

Anonymous. 1985. Special Series 1985: Europe's Machinery Industries - Report No. 2: New Technology and Innovation. *European Trends*(2): 38 - 53.

Archibugi D, Simonetti R. 1998. Objects and Subjects in Technological Interdependence. Towards a framework to monitor innovation. *International Journal of the Economics of Business* **5**(3): 295 - 309.

Arnold HM. 2001. Can great companies survive technology shocks? - A literature overview. *Munich School of Management, Electronic Working Paper Series*(1/2001)

Arnold HM. 2002. The next wave - trends and future technological changes in the machine tool industry: 1-35. INNOtec - Institute for Innovation Research and Technology Management: Munich.

Arnott D. 1983. The British Machine Tool Trauma. *Management Today*: 72 - 80.

Arthur B. 1988. Competing Technologies. In G Dosi (Ed.), *Technical Change and Economic Theory*: 590-607. Columbia University Press: New York.

Ashburn A. 1972-2000. Blue Bulletin. AMT - The Association for Manufacturing Technology, American Machinist

Astley GW. 1985. The two ecologies: Population and Community Perspectives on Organizational Evolution. *Administrative Science Quarterly*(30): 224-241.

Baba Y. 1989. The Dynamics of Continuous Innovation in Scale-Intensive Industries. *Strategic Management Journal* **10**(1): 89-100.

Bachmann D, Elfrink J, Vazzana G. 1996. Tracking the Progress of E-mail vs. Snail-mail. *Marketing Research* **8**

Baker NR, Green SG, Bean AS. 1985. How Management can Influence the Generation of Ideas. *Research Management* **28**(6): 35-43.

Baker NR, Green SG, Bean AS. 1986. Why R&D Projects Succeed or Fail. *Research Management* **29**(6): 29-35.

Baldwin WL, Scott JT. 1987. *Market Structure and Technological Change*. Harwood Academic Publishers: New York.

Bane B. 2000. Personal Interview. Vice President, Head of Telecommunications Practice, Mercer Management Consulting

Barnett WP. 1990. The Organizational Ecology of a Technological System. *Administrative Science Quarterly* **35**(1): 31-60.

Bass FM. 1969. A New-Product Growth Model for Consumer Durables. *Management Science* **15**(January): 215-227.

Bayus BL. 1998. An Analysis of Product Lifetimes in a Technologically Dynamic Industry. *Management Science* **44**(6): 763-775.

BCG. 1990. Europastudie für ausgewählte Zielländer - Dokumentarband. Verein Deutscher Werkzeugmaschinenfabriken e.V. -VDW

BCG-VDW. 1981. Japanstudie. The Boston Consulting Group, Verein Deutscher Werkzeugmaschinenfabriken e.V. (VDW): Frankfurt, München.

BfA. 1979. Marktinformation. Bundesstelle für Aussenhandelsinformation: Köln.

Bollen NPB. 1999. Real Options and Product Life Cycles. *Management Science* **45**(5): 670-684.

Bourdieu P. 1984. *Distinction: A social Critique of the Judgement of Taste* (R Nice, Trans.). Harvard University Press: Cambridge.

Bourgeois LJ, III, Eisenhardt KM. 1988. Strategic Decision Processes in High Velocity Environments: Four Cases in the Microcomputer Industry. *Management Science* **34**(7): 816-835.

Bower JL, Christensen CM. 1995. Disruptive Technologies: Catching the Wave. *Harvard Business Review* **73**(1): 43-53.

Brödner P, Schultetus W. 1992. Erfolgsfaktoren des japanischen Werkzeugmaschinenbaus - Eine Analyse des RKW, der IG Metall un des Vereins Deutscher Werkzeugmaschinenfabriken (VDW). Institut Arbeit und Technik des Wissenschaftszentrums NRW, Institut für angewandte Arbeitswissenschaft e.V.: Gelsenkirchen, Köln.

Brown JW, Utterback JM. 1985. Uncertainty and Technical Communications Patterns. *Management Science* **31**(3): 301-311.

Bühl A, Zöfel P. 2000. *SPSS Version 10* (10 ed.). Addison Wesley: München.

Bunnel D. 2000. *Making the Cisco Connection - The Story behind the real Internet Superpower* (1 ed.). John Wiley & Sons, Inc.: New York, Chichester, Weinheim, Brisbane, Singapore, Toronto.

Burns T, Stalker GM. 1966. *The Management of Innovation* (2 ed.). Tavistock: London.

Carroll GR, Hannan MT. 1989. Density Delay in The Evolution of Organizational Populations: A Model and Five Empirical Tests. *Admistrative Science Quarterly* **34**: 411-430.

Ceruzzi P. 1996. From Scientific Instrument to Everyday Appliance: The Emergence of Personal Computers, 1970 - 77. *History and Technology* **13**: 1-31.

Chakrabarti AK. 1974. The Role of Champion in Product Innovation. *California Management Review* **17**(2): 58-62.

Chalmers RE. 1998. A Journey to Japan. *Manufacturing Engineering* **121**(3): 80 - 85.

Chanaron JJ. 1999. Technological Change and Organization. *R & D Management* **29**(4): 425-426.

Christensen CM. 1993. The Rigid Disk Drive Industry: A History of Commercial and Technological Turbulence. *Business History Review* **67**(4): 531-558.

Christensen CM. 1997. *The Innovator's Dilemma: When New Technologies Cause Great Firms To Fail* (1 ed.). Harvard Business School Press: Boston.

Christensen CM, Bower JL. 1996. Customer power, Strategic Investment, and the Failure of Leading Firms. *Strategic Management Journal* **17**(3): 197-218.

Christensen CM, Suarez FF, Utterback JM. 1998. Strategies for Survival in Fast-changing Industries. *Management Science* **44**(12 (Part 2)): 207-220.

Christensen CM, Tedlow RS. 2000. Patterns of Disruption in Retailing. *Harvard Business Review*: 42-45.

Chrubasik B, Zimmermann H-J. 1987. Evaluierung der Modelle zur Bestimmung strategischer Schlüsselfaktoren. *Die Betriebswirtschaft* **47**(4): 426-450.

Cohen WM, Levin RC. 1989. Empirical Studies of Innovation and Market Structure. In R Schmalensee, RD Willig (Eds.), *Handbook of Industrial Organization*. North Holland: New York.

Cohen WM, Levinthal DA. 1990. Absorptive Capacity: A New Perspective on Learning and Innovation. *Administrative Science Quarterly* **35**(1): 128-152.

Cohen WM, Levinthal DA. 1994. Fortune Favors the Prepared Firm. *Management Science* **40**(2): 227-251.

Cook TD, Campbell DT. 1979. Quasi-Experimentation: chapter 3: Boston.

Coomber R. 1997. Using the Internet for Survey Research. *Sociological Research Online*: http://www.socresonline.org.uk/2/2/2.html.

Cooper AC, Schendel D. 1976. Strategic Responses to Technological Threats. *Business Horizons*(February): 61-69.

Cooper AC, Smith CC. 1992. How Established Firms Respond to Threatening Technologies. *Academy of Management Executive* **6**(2): 55-70.

Cooper RG. 1979. Identifying Industrial New Product Success: Project NewProd. *Industrial Marketing Management* **8**(2): 124-135.

Cooper RG, Kleinschmidt EJ. 1986. An Investigation into the New Product Process: Steps, Deficiencies, and Impact. *Journal of Product Innovation Management* **3**(2): 71-85.

Corsten H. 1982. *Der nationale Technologietransfer*: Berlin.

Cox D, Oakes D. 1984. *Analysis of Survival Data*. Chapman & Hall: New York.

Cox WJ. 1967. Product Life Cycles as Marketing Models. *The Journal of Business* **40**(4): 375-384.

Cyert RM, March JG. 1963. *A Behavioral Theory of the Firm*. Prentice Hall: Englewood Cliffs, NJ.

Daly A, Jones DT. 1980. The Machine Tool Industry in Britain, Germany, and the United States. *National Institute Economic Review*(92): 53.

D'Aveni RA, MacMillan IC. 1990. Crisis and the Content of Managerial Communications: A Study of the Focus of Attention of Top Managers in Surviving and Failing Firms. *Administrative Science Quarterly* **35**(4): 634-657.

David PA. 1985. Clio and the Economics of QWERTY. *American Economic Review* **75**: 332-337.

De Brentani U. 1989. Success and Failure in New Industrial Services. *Journal of Product Innovation Management* **6**(4): 239-258.

de Figueiredo JM, Kyle MK. 2000. Surviving the Gales of Creative Destruction: The Patterns of Innovative Activity in the Desktop Laser Printer Industry. *MIT Sloan Working Paper*

DeJong CA. 1998. All Machine Tools are not Created Equal. *Automotive Manufacturing & Production* **110**(3): 78 - 81.

Dillman DA. 1978. The Advantages and Disadvantages of Mail, Telephone, and Face-to-Face Surveys, *Mail and Telephone Surveys - The Total Design Method*: 39-78. John Wiley&Sons: New York, Chichester, Brisbane, Toronto, Singapore.

DiMaggio PJ, Powell WW. 1983. The Iron Cage Revisited: Institutional Isomorphism and Collective Rationality in Organizational Fields. *American Sociological Review* **48**: 147-160.

Dreher C. 1997. *Technologiepolitik und Technikdiffusion: Auswahl und Einsatz von Förderinstrumenten am Beispiel der Fertigungstechnik.* Nomos Verlagsgesellschaft: Baden-Baden.

Easingwood CJ. 1988. Product Lifecycle Patterns for New Industrial Products. *R&D Management* **18**(1): 23-32.

Ebadi YM, Utterback JM. 1984. The Effects of Communication on Technological Innovation. *Management Science* **30**(5): 572-586.

Ehrnberg E, Jacobsson S. 1997. Indicators of Discontinuous Technological Change: An Exploratory Study of two Discontinuities in the Machine Tool Industry. *R & D Management* **27**(2): 107-126.

Eichenbaum M. 1991. Technology Shocks and the Business Cycle. *Economic Perspectives (Federal Bank Reserve of Chicago)* **15**(2): 14-31.

EMO. 2001. *EMO Hannover - The World of Machine Tools - Catalogue.* Generalkommissariat der 14. EMO Hannover, Messegelände: Hannover.

Erdos PL. 1993. *Professional Mail Surveys*. Rober E. Krieger Publishing Co.: Malabar.

Ettlie JE. 1983. A Note on the Relationship between Managerial Change Values, Innovative Intentions and Innovative Technology Outcomes. *R&D Management* **13**(4): 231-245.

Eversheim W, König W, Weck M, Pfeifer T. 1980. *Anschriften der Freunde und ehemaligen Mitarbeiter des Laboratoriums für Werkzeugmaschinen und Betriebslehre der Rheinisch-Westfälischen Technischen Hochschule Aachen*: Aachen.

Fawcett CW, Roman D. 1976. Industry Overview for the Purchaser of Machine Tools. *Journal of Purchasing and Materials Management, Fall 1976*: 8-14.

Feigenbaum J. 1980. Innovation in the Factory. *IEEE Spectrum* **17**(1): 54.

FhG-ISI I, IWF. 1982. *Der Einsatz flexibler Fertigungssysteme - Technische, einführungsorganisatorische, wirtschaftliche und arbeitsplatzbezogene Aspekte*: Karlsruhe.

Fienberg S. 1977. *The Analysis of Cross-classified categorical data.* The MIT Press: Cambridge, Massachusetts, London, England.

Fleischer M. 1997. *The Inefficiency Trap: Strategy Failure in the German Machine Tool Industry.* Wissenschaftszentrum Berlin für Sozialforschung - WZB: Berlin.

Forrant R. 1997. Good jobs and the Cutting Edge: The U.S. Machine Tool Industry and Sustainable Prosperity. *Working Paper - University of Massachusetts Lowell*(199): 1 - 30.

Foster RN. 1986. *Innovation. The Attacker's Advantage.* Summit Books: New York.

Frederickson JW, Mitchell TR. 1984. Strategic Decision Processes: Comprehensiveness and Performance in an Industry with an Unstable Environment. *Academy of Management Journal* **27**(2): 399-423.

Gaitanides M, Wicher H. 1985. Venture Management: Strategien und Strukturen der Unternehmensentwicklung. *DBW*: 414-426.

Gardner D. 1987. The Product Life Cycle. *Die Unternehmung* **41**(3): 219-231.

Gatignon H, Robertson TS. 1985. A Propositional Inventory for New Diffusion Research. *Journal of Consumer Research* **11**: 849-867.

Gebhart J. 1995. Waves of Change. *Sloan Management Review* **36**(3): 103-104.

Gilbert R, Newbery D. 1984. Preemptive Patenting and the Persistence of Monopoly: Comment. *American Economic Review* **74**: 238-242.

Gilbert RJ, Newberry DM. 1982. Preemptive Patenting and the Persistence of Monopoly. *American Economic Review* **72**: 514-526.

Glasmeier A. 1991. Technological Discontinuities and Flexible Production Networks: The Case of Switzerland and the World Watch Industry. *Research Policy* **20**(5): 469-485.

Glunk FR. 1989. *100 Jahre Pittler 1889 - 1989.* Britting Verlag: Langen.

Goldhar JD. 1994. Mastering the Dynamics of Innovation: How Companies Can Seize Opportunities in the Face of Technological Change. *Sloan Management Review* **35**(4): 97-98.

Greenstein SM, Wade JB. 1998. The Product Life Cycle in the Commercial Mainframe Computer Market, 1968-1982. *RAND Journal of Economics* **29**(4): 772-789.

Griliches Z. 1957. Hybrid Corn: An Exploration in the Economies of Technological Change. *Econometrica* **25**: 501-522.

Griliches Z. 1979. Issues in Assessing the Contribution of Research and Development to Productivity Growth. *Bell Journal of Economics* **10**: 92-116.

Hannan MT, Freeman J. 1977. The Population Ecology of Organizations. *American Journal of Sociology* **82**: 929-064.

Harabi N. 1998. Innovation through vertical relations between firms, suppliers and customers: A study of German firms. *Industry and Innovation* **5**(2): 157 - 172.

Harhoff D. 1991. Strategic Spillover Production, Vertical Organization, and Incentives for Research and Development, *Sloan School of Management*. Massachusetts Institute of Technology: Cambridge, Mass.

Harrop J. 1985. Crisis in the Machine Tool Industry: A Policy Dilemma for the European Community. *Journal of Common Market Studies* **24**(1): 61 - 76.

Hatfield DE, Tegarden LF, Echols AE. 2001. Facing the Uncertain Environment from Technological Discontinuities - Hedging as a Technology Strategy. *Journal of High Technology Management Research* **12**: 63 - 76.

Hauschildt J. 1997. *Innovationsmanagement* (2 ed.). Vahlen: München.

Henderson R. 1993. Underinvestment and Incompetence as Reponses to Radical Innovation: Evidence from the Photolithographic Alignment Equipment Industry. *RAND Journal of Economics* **24**(2): 248-270.

Henderson RM, Clark KB. 1990. Architectural Innovation: The Reconfiguration of Existing Product Technologies and the Failure of Established Firms. *Administrative Science Quarterly* **35**(1): 9-30.

Hildebrand DK. 1986. *Statistical Thinking for Behavioral Scientists*. Duxbury Press: Boston.

Hill CT, Utterback JM. 1980. The Dynamics of Product and Process Innovation. *Management Review* **69**(1): 14-20.

Hoffmann K. 1972. *Der Produktlebenszyklus - Eine kritische Analyse.* Verlag Rombach Freiburg: Freiburg i. Breisgau.

Hollander S. 1964. *The Sources of Increased Efficiency: A Study of Du-Pont Rayon Plants*. The M.I.T. Press: Cambride, MA.

Hsu S, Chang P-L. 2000. Promoting Technological Capabilities of Small and Medium-Sized Enterprises through Industry-University Cooperation: Case Study of Taiwan Machine Tool Industry. *International Journal of Manufacturing Technology and Management* **1**(2,3): 257 - 270.

Industry-Expert-Interview-Series. 2000-2001. Interview Series

Johne FA. 1984. How Experienced Product Innovators Organize. *Journal of Product Innovation Management* **1**(4): 210-223.

Jones DE, Port O. 1984. The King of Knockoffs Rushes to Go from Imitation to Innovation, *Business Week*: 188 - 192

Kanuk L, Berenson C. 1995. Mail surveys and response rates: A literature review in Journal of Marketing Research. *Journal of the Market Research Society* **17**

Katz R, Allen TJ. 1982. Investigating the Not-Invented-Here (NIH)-Syndrome. *R&D Management*: 7-19.

Kawakami K, Abe S. 1985. Machine Tools Build Prosperity for Related Industries/ Machine Tool Industry in Japan Today. *Business Japan* **30**(8): 89 - 97.

Kennedy P. 1996. Qualitative Dependent Variables, *A Guide to Econometrics*, 4 ed.: 233-248. Blackwell Publishers: Oxford.

Klepper S. 1996. Entry, Exit, Growth, and Innovation over the Product Life Cycle. *American Economic Review* **86**(3): 562-583.

Klepper S. 1999. Firm Survival and the Evolution of Oligopoly. *Carnegie Mellon University Working Paper*

Klepper S, Graddy E. 1990. The Evolution of New Industries and the Determinants of Market Structure. *Rand Journal of Economics* **21**(1): 27-44.

Klump R, Männel B. 1995. "Lange Wellen" der Unternehmenskonjunktur. *Zeitschrift für Unternehmensgeschichte* **40**(1): 1 - 34.

Kortge GD, Okonkwo PA, Burley JR, Kortge JD. 1994. Linking Experience, Product Life Cycle, and Learning Curves. *Industrial Marketing Management* **23**: 221-228.

Kotler P, Bliemel F. 1995. Marketingstrategien für die Phasen im Produkt-Lebenszyklus, *Marketing Management*, 8 ed.: 557-596: Stuttgart.

Kuba Y. 1989. *Meister der Fertigungstechnologie - Die 70jährige Geschichte von Mazak*. Yamazaki Mazak Corp.: Taito-ku, Tokyo.

Kulp D, Hunter R. 1999. Pro and Con - Internet Interviewing. *Marketing Research*(33)

Laux H. 1998. *Entscheidungstheorie* (4. Auflage ed.): Berlin.

Lee H-H, Stone JA. 1994. Product and Process Innovation in the Product Life Cycle: Estimates for U.S. Manufacturing Industries. *Southern Economic Journal* **60**(3): 754-763.

Lee J-R, O Neal DE, Pruett MW, Thomas H. 1995. Planning for Dominance: A Strategic Perspective on the Emergence of a Dominant Design. *R & D Management* **25**(1): 3-15.

Lee KR. 1996. The role of user firms in the innovation of machine tools: The Japanese case. *Research Policy* **25**: 491-507.

Leonard-Barton D. 1992. Core Capabilities and Core Rigidities: A Paradox in Managing New Product Development. *Strategic Management Journal* **13**: 111-125.

Levitt T. 1960. Marketing Myopia. *Harvard Business Review*(July-August): 45-56.

Lindblom CE. 1959. The Science of 'Muddling' Through. *Public Administration Review* **19**: 79-88.

Link PL. 1987. Keys to New Product Success or Failure. *Industrial Marketing Management* **16**(2): 109-119.

Loch CH, Huberman BA. 1999. A Punctuated-Equilibrium Model of Technology Diffusion. *Management Science* **45**(2): 160-177.

Mansfield E. 1983. Technological Change and Market Structure: An Empirical Study. *American Economic Review Proceedings* **73**: 205-209.

Martino JP. 1999. The Environment for Technological Change. *International Journal of Technology Management* **18**(1,2): 4-10.

Marx TG. 1979. Technological Change in the Structure of the Machine Tool Industry. *MSU Business Topics* **27**(1): 41-47.

Mazzoleni R. 1997. Learning and Path-Dependence in the Diffusion of Innovations: Comparative Evidence on Numerically Controlled Machine Tools. *Research Policy* **26**(4,5): 405 - 428.

Mazzoleni R. 1998. The Agency System in the International Distribution of U.S. Machine Tools, 1900 - 1915: Social Norms and Contracts. *Business and Economic History* **27**(2): 420 - 430.

Mensch G. 1979. *Stalemate in Technology*. Ballinger: Cambridge, MA.

Mercer-UBM. 1992a. Company Profiles. Mercer Management Consulting, Unternehmensberatung München: Berlin, München.

Mercer-UBM. 1992b. Competence Presentation Machine Tools. Mercer Management Consulting/ Unternehmensberatung München: Munich.

Merchant EM. 2000. Manufacturing - Yesterday, Today and Tomorrow. *Working Paper - Institute of Advanced Manufacturing Sciences*

Miller D, Friesen PH. 1984. A Longitudinal Study of the Corporate Life Cycle. *Management Science* **30**(10): 1161-1183.

Mitchell W. 1989. Whether and When? Probability and Timing of Incumbents' Entry into Emerging Industrial Subfields. *Administrative Science Quarterly* **34**(2): 208-230.

Mitchell W, Singh K. 1996. Survival of Businesses Using Collaborative Relationships to Commercialize Complex Goods. *Strategic Management Journal* **17**: 169-195.

Mittag H. 1985. *Technologiemarketing*: Bochum.

Mueller DC. 1997. First-Mover Advantages and Path Dependence. *International Journal of Industrial Organization* **15**: 827-850.

Mueller S, Kesselmann P. 1996. Akzeptanz von computergestützten Erhebungsverfahren – Ein empirischer Vergleich mit der traditionellen Fragebogentechnik. *Marketing*(3): 191-202.

Myers S, Sweezy EE. 1978. Why Innovations Fail. *Technology Review* **March/April**: 41-46.

Nadler DA, Tushman ML. 1979. A Congruence Model for Diagnosing Organization Behavior. In D Kolb, J Rubin, J McIntyre (Eds.), *Organizational Psychology: A book of readings*, 3rd edition ed. Prentice Hall: Englewood Cliffs, NJ.

National-Research-Council. 1987. Management of Technology: The Hidden Competitive Advantage. National Academy Press: Washington, D. C.

Nelson RR, Winter SG. 1982. *An Evolutionary Theory of Economic Change*. Harvard University Press: Cambridge, Massachusetts.

Norton JA, Bass FM. 1987. A Diffusion Theory Model of Adoption and Substitution for Successive Generations of High-Technology Products. *Management Science* **33**(9): 1069-1086.

Paolillo JG, Brown WB. 1978. How Organizational Factors Affect R&D Innovation. *Research Management* **21**(2): 12-15.

Pfeffer J, Salancik GR. 1978. *The External Control of Organizations: A Resource Dependence Perspective*. Harper & Row: New York.

Pfeiffer R, Lohmüller B. 1999. Vergleich betrieblicher Leistungen für Unternehmen des Sektors Maschinenbau aus Großbritannien. *zfo - Zeitschrift für Organisation* **5**: 260-264.

Pisano GP. 1990. The R&D Boundaries of the Firm: An Empirical Analysis. *Administrative Science Quarterly* **35**(1): 153-176.

Pisano GP, Wheelwright SC. 1995. The New Logic in High-Tech R&D. *Harvard Business Review* **Sept.-Oct.**: 93-105.

Pohlmann G. 1992. Die standardisierte Befragung als Kommunikationssituation. *Planung und Analyse*(4): 34-40.

Polli R, Cook V. 1969. Validity of the Product Life Cycle. *The Journal of Business* **42**(4): 385-400.

Porter ME. 1980. *Competitive Strategy*. Free Press: New York.

Price RM. 1996. Executive Forum: Technology and Strategic Advantage. *California Management Review* **38**(3): 38-56.

Production M. 1983. *The Fifth Survey of Machine Tools and Production Equipment in Britain*. Morgan Grampian: London.

Quinn JB. 1978. Strategic Change: Logical Incrementalism. *Sloan Management Review* **20**(1): 7-21.

Quinn JB. 1980. *Startegies for Change*. Dow-Jones Irwin: Homewood, IL.

Raiser E. 2001. Wie das erste numerisch gesteuerte Bearbeitungszentrum bei Burkhardt+Weber entstand, 1 ed. Burkhardt+Weber Gruppe

Rao AS. 1997. On the contribution of technology shocks to business cycles. *Federal Reserve Bank of Minneapolis Quarterly Review* **21**(3): 15-27.

Rauleder S. 1986. Methoden zur Erarbeitung von Wettbewerbsvorteilen - am Beispiel des deutschen Werkzeugmaschinenbaus, *Fachbereich Betriebswirtschaft*. Ludwig-Maximilians-Universität: München.

Reidenbach RE, Moak DL. 1986. Exploring Retail Bank Performance and New Product Development: A Profile of Industry Practices. *Journal of Product Innovation Management* **3**(3): 187-194.

Reinganum J. 1983. Uncertain Innovation and the Persistence of Monopoly. *American Economic Review* **73**: 741-748.

Reintjes FJ. 1991. *Numerical Control: Making a New Technology*. Oxford University Press: New York, Oxford.

Rendeiro JO. 1985. How the Japanese Came to Dominate the Machine Tool Business. *Long Range Planning* **18**(3): 62-67.

Rifkin G. 1994. The Myth of Short Life Cycles. *Harvard Business Review* **72**(4): 11.

Rink DR, Swan JE. 1979. Product Life Cycle Research: A Literature Review. *Journal of Business Research* **7**: 219-242.

Romeo AA. 1975. Interindustry and Interfirm Differences in the Rate of Diffusion of an Innovation. *The Review of Economics and Statistics* **57**(3): 311-319.

Rubenstein AH, Chakrabarti AK, O'Keefe RD, Souder WE, Young HC. 1976. Factors Influencing Innovation Success at the Project Level. *Research Management* **19**: 15-20.

Saxanian A. 1995. *The Regional Advantage*. Harvard University Press: Cambridge, MA.

Schewe G. 1994. Erfolg im Technologiemanagement - Eine empirische Analyse der Imitationsstrategie. *Zeitschrift für Betriebswirtschaft* **64**(8): 999-1026.

Schmookler J. 1962. Economic Sources of Inventive Activity. *The Journal of Economic History* **23**(1): 1 - 20.

Schnell R, Hill P, Esser E. 1995. *Methoden der empirischen Sozialforschung*: München.

Schrader S. 1995. Informal Alliances: Information Trading between Firms. *Advances in Global High-Technology Management* **5**(B): 31-55.

Schrader S. 1996. Innovationsmanagement. In W Kern, H-H Schröder (Eds.), *Handwörterbuch der Produktionswirtschaft*: 744-758. Schäffer-Poeschel Verlag: Stuttgart.

Schröder S. 1995. *Innovation in der Produktion: eine Fallstudienuntersuchung zur Entwicklung der numerischen Steuerung*. Carl Hanser Verlag: München, Wien.

Schumpeter J. 1912. *Theorie der Wirtschaftlichen Entwicklung*: Leipzig.

Schumpeter JA. 1934. *The Theory of Economic Development*. Harvard University Press: Cambridge, MA.

Schumpeter JA. 1939. *Business Cycles*

Schumpeter JA. 1942. Capitalism, Socialism and Democracy, *Capitalism, Socialism and Democracy*: 63-86. Harper: New York.

Schwab G. 1996. Die Entwicklung der deutschen Werkzeugmaschinenindustrie von 1945 - 1995: 1 - 62. Verein Deutscher Werkzeugmaschinenfabriken e.V.: Frankfurt am Main.

Sciberras E, Payne BD. 1985. *Machine Tool Industry* (1 ed.). Longman Group Limited: London.

Shane S. forthcoming. Prior Knowledge and the Discovery of Entrepreneurial Opportunities. *Organization Science*

Shapiro C, Varian HR. 1999. The Art of Standard Wars. *California Management Review* **41**(2): 8-32.

Shea J. 1998. What do Technology Shocks do? *NBER Working Paper Series* **6632**: 1-58.

Sheehan KB, Grubbs Hoy M. 1999. Using e-mail to survey Internet users in the United States: Methodology and Assessment. *Journal of Computer-Mediated Communication* **4**(3): http://jcmc.huji.ac.il/vol4/issue3/sheehan.html.

Siegwart H, Senti R. 1995. *Product Life Cycle Management - Die Gestaltung eines integrierten Produktlebenszyklus*. Schäffer-Poeschel Verlag: Stuttgart.

Simon HA. 1957. *Administrative Behavior*. The Free Press: New York.

Smith CG, Cooper AC. 1988. Established Companies Diversifying into Young Industries: A Comparison of Firms with Different Levels of Performance. *Strategic Management Journal* **9**: 111-121.

Spur G. 1991. *Vom Wandel der industriellen Welt durch Werkzeugmaschinen*. Carl Hanser Verlag: München, Wien.

Spur G. 1996. *Die Genauigkeit von Maschinen: eine Konstruktionslehre* (1 ed.). Carl Hanser Verlag: München, Wien.

Spur G, Specht D, Schröder S. 1994. Case B: Die Numerische Steuerung (NC-Machine Tools). In TAoSaTiBWGCaT Innovation) (Ed.), *Culture and Technical Innovation: a Cross-Cultural Analysis and Policy Recommendations*, Vol. 9: 621 - 735. Walter de Gruyter: Berlin.

Steeds W. 1969. *A History of Machine Tools 1700 - 1910*. The Clarendon Press: Oxford.

Steele T, Schwending L, Kilpatrick J. 1992. Duplicate responses to multiple survey mailings: A problem? *Journal of Advertising Research* **32**

Stewart B. 1999. Multimedia Market Research - Digital technology presents new opportunities for product studies. *Marketing Research*(34)

Suarez FF, Utterback JM. 1995. Dominant Designs and the Survival of Firms. *Strategic Management Journal* **16**(6): 415-430.

Teece D, Pisano G. 1994. The Dynamic Capabilities of Firms: an Introduction. *Industrial and Corporate Change* **3**(3): 537-556.

Theobald A, Brabänder E. 1998. Möglichkeiten der Sekundär- und Primärmarktforschung im Internet. *Kaiserslauterer Schriftenreihe Marketing* **6**

Tichy G. 1991. The Product-Cycle Revisited: Some Extensions and Clarifications. *Zeitschrift für Wirtschafts- und Sozialwissenschaften* **111**: 27-54.

Tilton JE. 1971. *International Diffusion of Technology: The Case of Semiconductors*. The Brookings Institution: Washington, D.C.

Tirole J. 1988. Research and Development and the Adoption of New Technologies, *The Theory of Industrial Organization*: 389-421. The MIT Press: Cambridge, Massachusetts.

Toutenburg H. 2000. *Induktive Statistik* (2nd ed.). Springer: Berlin, Heidelberg.

Tschätsch H, Charchut W. 1991. *Werkzeugmaschinen: Einführung in die Fertigungsmaschinen der spanlosen und spanenden Formgebung* (6 ed.). Carl Hanser Verlag: München, Wien.

Tsurumi H, Tsurumi Y. 1980. A Bayesian Test of the Product Life Cycle Hypothesis as Applied to the US Demand for Color-TV Sets. *International Economic Review* **21**(3): 583-597.

Tushman ML, Anderson P. 1986. Technological Discontinuities and Organizational Environments. *Administrative Science Quarterly* **31**(3): 439-465.

Unger H. 1985. Industrial America: Strategies for Recovery in Machine Tool Industry. *Industrial Management* **9**(4): 13.

USDoC. 1959 - 1968. Census. US Department of Commerce

USITC. 1983. Competitive Assessment of the US Metalworking Machine Tool Industry. United States International Trade Commission: Washington, D.C.

Utterback JM. 1994. *Mastering the Dynamics of Innovation*. Harvard Business School Press: Boston, MA.

VDMA. 1993. Kennzahlenkompass. Verband deutscher Maschinen- und Anlagenbau e.V.: Frankfurt am Main.

VDW. 1974. Die deutsche Werkzeugmaschinen-Produktion in Zeitreihen und Trends 1958 - 1974. Fachgemeinschaft Werkzeugmaschinen im VDMA: Frankfurt am Main.

VDW. 1999a. The German Machine Tool Industry in 1999. Verein Deutscher Werkzeugmaschinenfabriken e.V. (VDW): Frankfurt am Main.

VDW. 1999b. The German Machine Tool Industry in 1999: 1 - 52. Verein Deutscher Werkzeugmaschinenfabriken e.V.: Frankfurt.

VDW. 2001. Historical Time Series of the German Machine Tool Industry. VDW German Machine Tool Builders' Association: Frankfurt.

VDW. 2002. The German Machine Tool Industry in 2001. Verein Deutscher Werkzeugmaschinenfabriken e.V. (VDW): Frankfurt am Main.

VDW WAMCiawI-I. Strategic Study on EC Machine Tool Sector: 43. VDW

von Hippel E. 1988. *The sources of innovation*: New York.

von Hippel E. 1994. 'Sticky Information' and the Locus of Problem Solving: Implications for Innovation. *Management Science* **40**(4): 429-439.

Wagoner HD. 1966. *The US Machine Tool Industry*. The MIT Press: Cambridge, MA.

Willard GE, Cooper AC. 1985. Survivors of Industry Shake-outs: the Case of the U.S. Color Television Set Industry. *Strategic Management Journal* **6**: 299-318.

Witte E. 1968a. Organisation komplexer Entscheidungsverläufe. *ZFBF* **20**(9): 581-599.

Witte E. 1968b. Phasen-Theorem und Organisation komplexer Entscheidungsverläufe. *ZFBF* **20**(9): 625-652.

Witte E. 1972. Das Informationsverhalten in Entscheidungsprozessen: 1-88: Tübingen.

Witte E. 1973. *Organisation für Innovationsentscheidungen*: Göttingen.

Witte H. 1991. *Werkzeugmaschinen: Grundlagen und Prinzipien in Aufbau, Funktion, Antrieb und Steuerung spangebender Werkzeugmaschinen* (7 ed.). Vogel Buchverlag: Würzburg.

Xunxian L. 1981. China's Machine Tool Industry. *The China Business Review* **8**(6): 30 - 36.

Yin RK. 1984. *Case Study Research: Design and Methods*. Sage Publishers: Beverly Hills.